# THE ESSENTIAL

### DISCOVERING WHAT REALLY MATTERS IN AN AGE OF DISTRACTION

## SOREN GORDHAMER

wisdom 2.0

THE ESSENTIAL: DISCOVERING WHAT REALLY MATTERS IN AN AGE OF DISTRACTION

© 2025 Soren Gordhamer

A Wisdom 2.0 Book

Print ISBN: 979-8-218-63839-9

Ebook ISBN: 979-8-218-63838-2

www.wisdom2summit.com

DISCLAIMER: This is book is drawn from real-life experiences. In a few places, I changed names or small details out of respect for the privacy of the people who were a part of the journey.

# PRAISE FOR SOREN GORDHAMER

"Profound and timely lessons in fine-tuning your relationship to yourself and what might be possible in your life were you to listen deeply to what is being called out of you, and why. A friendly reminder of your agency in navigating the times we live in with integrity and purpose, and ultimately, with wisdom."

—**Jon Kabat-Zinn**, Founder of MBSR, Author, *Full Catastrophe Living* and *Coming to Our Senses*

"A book that has the power to radically transform the reader's life. *The Essential* is a must read that will help people become a more compassionate version of themselves."

— **Yung Pueblo**, #1 NYT Bestselling Author

"Soren reminds us that the most profound answers are found not in doing more, but in being present for what matters most."

— **Arianna Huffington**, Founder & CEO, Thrive Global

"Here is wisdom in a delicious simple form, with reminders of how to stay present, true to ourselves and our hearts amidst it all."

—**Jack Kornfield**, Author, A *Path With Heart*

"A charming collection of teaching anecdotes from the author's diligent inner and outer explorations, helping us to remember what matters in life."

— **Gabor Maté, M.D.**, Author, *The Myth of Normal*

"In *The Essential*, Soren shares powerful insights from his own life to inspire and guide us on our own transformative journey."

— **Dr. Mark Hyman**, *New York Times* bestselling author of *Young Forever* and *The Pegan Diet*

"*The Essential* is a spiritual compass for anyone seeking greater peace and a deeper connection to what truly matters."

— **Deepak Chopra, M.D.**, Author of *Digital Dharma*

"A book with deep honesty and integrity. Soren's journey will open your eyes to what matters—and what doesn't—in your own life, and that is a profound gift."

— **Dr. Richard Schwartz**, Founder, Internal Family Systems

"Soren reminds us to not lose focus on what matters most: maintaining our shared humanity."

— **Jeff Weiner**, Executive Chairman, LinkedIn

"*The Essential* is an excellent handbook for modern living."

— **Sharon Salzberg**, Author of *Lovingkindness* and *Real Life*

"A clear and compassionate compendium of deep lessons for our distracted age. I learned so much from this book."

—**James Nestor,** Author of *Breath: The New Science of a Lost Art*

"A jewel of a book."

— **Roshi Joan Halifax**, Upaya Zen Center

"Part wisdom philosophy, part memoir, part stroll-through-the-park-with-an-ancient sage, this precious jewel of a book will help readers reflect on how they are living, what's driving them to make the decisions they make, and how to step into a more spacious way of being."

— **Elizabeth R. Koch**, Founder and CEO of Unlikely Collaborators

# CONTENTS

# PART I

## THE INTRODUCTION

In our world today, we know a lot …

We know stock prices up to the minute, the latest breaking
news, what all our friends on social media are doing …

But we often know much less about what actually matters.

What helps us feel alive?
What pain or trauma still lingers inside us?
What love in us wants to be acknowledged and expressed?

Something in us knows we have a choice:

Lose ourselves in the endless digital scroll, or look up and ask,
**what is actually essential?**

# THE OPPORTUNITY

T his book is not about Artificial Intelligence, but it was, in part, informed and inspired by it.

"I wish I could slow AI development, but I can't," admitted the CEO of one of the world's leading AI companies during a video call we had.

A few months earlier, I'd met and interviewed another prominent figure in AI: Sam Altman, CEO of OpenAI. Speaking with Sam backstage and interviewing him onstage, one thing became crystal clear: AI is advancing—and it's advancing very, very fast. This isn't just a wave on the ocean; it's a tsunami. As I learned more about AI, I found myself caught between two powerful reactions: "Wow, this is amazing!" and "Shit, this is scary."

I realized that advancements that I once thought would take decades might happen in just a few years. Every aspect of our society will be transformed. While some of this impact is up to the AI companies and the government, some is also up to us. AI will give us abilities and powers we never knew possible, and create enormous potential for positive change. The question isn't whether AI will advance. It will! The real question is, will humans advance?

Will we have the newest, coolest, most advanced AI … with the same old loneliness, hatred, greed, addiction, and despair? Or will we dare to create something profoundly new? For the latter, humanity itself must evolve.

As we navigate advancements in *artificial* intelligence, it seems like an opportune time to discover and harness our own innate *natural* intelligence, our natural wisdom. In fact, I would argue that perhaps the emerging AI is an invitation of sorts for humanity to better understand who we are and what really matters. And there is no time to lose. AI is advancing at lightning speed. Humanity needs to as well.

**On the outside, our world has changed massively. But on the inside, nothing much has changed.**

While AI is getting more intelligent by the day, our mind and body still work the same way they have for millennia. What eases suffering and creates happiness has not changed. No AI has yet mastered the secrets to happiness or fulfillment. At present, even with all of our gadgets and AI powers, we somehow feel more lonely, agitated, and disconnected than ever.

I have had the great pleasure to interview and speak in depth with some of the most successful and influential people of our age, including the founders of Meta, OpenAI, Anthropic, LinkedIn, and others. If life was an outward journey of accumulation, and not an inside voyage, they would know it. Yet each will tell you the same thing: outer accomplishments and fame are fine, but none of this satiates. None of this deepens a true connection with life. No amount of attention, applause, money, or power makes someone happy. There is another dimension.

Of course, there is nothing wrong with money, titles, and possessions. They are wonderful to have and enjoy, but if the only way we know ourselves is through our roles and possessions, we miss something quite important: a deeper intelligence inside ourselves, what some people call essence, presence, or awareness or Self. The greatest works of art, creative breakthroughs, and acts of generosity come from our ability to access an intelligence beyond our stories, roles, and identities.

**We have become enthralled with the analytical mind and the material world, while losing access to other dimensions, including the mystical and spiritual. And we are paying the price.**

This book is about knowing what is essential in life, connecting with who we are beyond the outer forms, and living aligned with what matters most. It is about tapping into an intelligence that has nothing to do with our age, our job title, the size of our bank account, or whether we are deemed successful or unsuccessful in the world. Thankfully, there is another dimension.

Since my teenage years, I have been driven by a deep yearning to understand the fundamental truths of existence. In this journey, I have had the great fortune to walk across numerous countries, to interview and get to know many of the great wisdom teachers of our age, and work with kids in New York City juvenile halls. And each group has its own wisdom.

This book is a reflection of my own learnings. You can read it sequentially or jump in at any point. The book is not a straight line; it bobs and weaves, and overlaps sometimes. It is written more as a creative flow. Take what resonates with you and leave the rest.

I wrote this book because if anything will help humanity survive and thrive in the years ahead, it may just be a radical shift inward, a spiritual renaissance, a remembering of who we are and what is essential. In a time that glorifies and high-lights the non-essential world of clicks and likes, and algo-rithms attempt to keep our attention glued to screens, the deeper dimensions of life are still here. There is still the inner domain.

There is still the essential.

# THE WORLD TODAY

E very day, billions of videos, posts, messages, and more are shared online. Amazing, right? Just about any information we could ever dream of likely exists in a video or social media post somewhere.

And every day, we keep adding more content. I mean, how many times have we seen a social media post on the importance of drinking water or exercising ... or of someone smiling at the beach or posting about a new job? And we just keep posting and watching the similar images and videos, day-after-day, night-after-night.

It's like having a giant catered buffet in front of us at all times, hanging from our neck, with new dishes continually appearing. As long as we have our phone with us, there's an endless stream of content whispering, "Here's a delightful morsel. Take a bite." Some of this content is nourishing and helpful, and some is pure junk, poisoning our thoughts just as certain foods poison our bodies.

As such, this is both the best of times and the worst of times. It is the best of times if we can manage the flow of information and harness the immense potential of AI with wisdom. Yet, it is the worst of times if we live trapped in a matrix of clicks and likes, guided by companies and algorithms that care little about our well-being. We then reach the end of our lives asking, "Where did my life go? I knew everything about what was going on outside of me, but almost nothing about what was inside me."

**Today, we have access to massive amounts of information, yet we are longing for something else, for wisdom.**

In this book, I attempt to share what is both sufficient and enough. My goal is not to overwhelm you with endless content or cloud your mind with complex theories. Instead, it is to share, in a simple and direct way, lessons I have learned along the way. It is to say just enough ... and to focus only on the essential.

Please read this book with your heart as well as your mind.
The point is not for everything to make sense.

"There is a voice inside that doesn't use words. Listen."
— Rumi

# INNER POWER

I am a passionate AI enthusiast and user, as well as an investor in various AI companies. I'm excited about the potential to harness AI across many fields—from healthcare to education—to create meaningful impact. However, while AI is amazing, it is not the answer to all of our personal or global challenges. Yes, AI is smarter than humans in many ways and is getting smarter, but AI will never surpass the depth of consciousness that underlies humanity and all life. There is an intelligence that is inherent in humans, and if we lose access to this, we are doomed.

Some years ago, I was in the home of one of the Bay Area's most notable tech leaders, sitting with about 30 other industry figures. These were the founders and CEOs of some of tech's most influential and profitable companies. We were there for an afternoon with the late Vietnamese Buddhist monk, Thich Nhat Hanh. Thay, as his students referred to him, sat before us in the living room, calmly speaking about meditation and mindfulness. The room was completely silent, with no phones visible. Thay spoke about the practice of meditation, mindfulness, and ethical living.

As I looked around the room at these leaders who wielded incredible power in the world, I asked myself, "Why are they here? Why have they cleared their schedules to listen to this simple monk?" Gradually, it became clear to me who truly held the power in that room—and it wasn't the tech leaders.

These leaders ran billion-dollar companies, had tens of thousands of employees, created products used across the globe, and had access to all the coolest technologies, but they showed up because they were missing something. They had one kind of power, but they knew very little about this other kind of power, what we may call inner power. And they arrived this afternoon to see if this monk might help them access it.

**Outward power is easily seen, inner power is easily felt.**

Thich Nhat Hanh was a Buddhist monk and social activist, who had no money, no possessions, and had never built a business. He didn't know anything about the stock market, hiring or firing employees, or bringing a product to market. He knew, however, what was essential: he knew how to work with life and find peace. He practiced sitting and walking meditation, day after day, observing life as it is. This was his power, and everyone there knew it.

We think worldly power is key, and it matters somewhat, but there is also the power of presence, of Being, of knowing who we are. He had a power no one else in that room had: the power to sit by himself and be at peace.

There's a story of a samurai who made his way across the land, killing anyone in his path. As he approached a village, he saw an old monk sitting calmly outside a temple.

Furious that the monk didn't flee, the samurai shouted, "Don't you know who I am? I could kill you in an instant without batting an eye."

The monk replied, "And don't you know who I am? I can sit here and have you kill me without batting an eye."

At that moment, the samurai understood a different kind of power.

There is a deeper intelligence inside us, and if we do not know it, no new technology can save us. For this intelligence, there is no need for servers, chips, data centers, or high speed internet. It is accessible to each one of us, and it costs nothing.

How do we harness it? Well, let's explore that together.

Real power is the thing that no one can take away.

# THE GAME

A massive effort is underway to encourage us to forget what truly matters.

For many of us living in developed countries, there is a game we are playing, and the rules are very important to understand. It is not necessarily a bad game or a good game, but not knowing the rules puts us at a huge disadvantage. This game is not played on grassy fields or clay tennis courts or basketball gyms; it is instead played on various digital platforms. The goal of the game is simple: get and keep our attention.

**Some of the greatest minds of our generation, armed with the most advanced technologies and billions of dollars, are focused on one goal: capturing and holding our attention.**

Every age presents unique challenges and opportunities. Historically, there were times of famines, times of war, and times of bounty. Our current time comes with its share, but one that affects almost all of us is the influence of screens. Today's screens offer incredible entertainment and learning.

The joy of watching a hilarious video from a person in a different part of the world can be beautiful. At the same time, social technologies are impacting our attention in dramatic ways ... and this is not accidental.

We have supercomputers with the most advanced AI pointing at our brains, directing us almost at will.

**Evan Williams, cofounder of Twitter (now X), may have said it best at Wisdom 2.0 some years ago, "Our attention has been hijacked and it's primarily been hijacked by advertising-driven business models that are not driven to make us more wise."**

Now, you might counter, *"Hold it, technology is a part of our lives, we all have to use it. I want to learn how to play the game so I can expand my social media following, amplify my voice in the world, grow my business, build my brand, and increase my income."*

Agreed. It's not about resisting it, but about understanding and consciously engaging with it. Social media can be a powerful tool for both spreading a message and entertainment.

Or you may say, *"I want to know the game, so I am not as consumed by social media, and I can better make time for myself, my real friends, and other things I care about."*

That is awesome too.

Or, *"I spend an average of three to four hours a day scrolling on my phone, and I can tell you almost nothing about that time. It is like a daze."*

The last one, of course, is the main concern.

For the vast majority of us, technology is essential in our lives ... and it also helps to know how the game is played so we can use it instead of being used by it.

I know many of the people who built the game, and it plays out in a number of ways. There are sites and platforms, of course, that are not geared to keep and sustain our attention. There are others, however, where the entire game is to keep us hooked. Every post, video or ad we see is not random; it is based on data the platform company has gathered from all our activity. The AI algorithm then presents the content it thinks will most likely keep us on the platform.

If the content that keeps us on the platform the most makes us feel unworthy, the algorithm will feed us more content that increases our sense of unworthiness. If greed keeps us hooked, we will be served content that fuels our desire for more. If it is humor, humor we will get. The algorithm does not really care. More time = more ads = more money. I'm not saying this is all bad, it is just the game.

**The algorithms try to convince us that everything non-essential in life (our status, what a celebrity just did) is essential, and everything actually essential (time with friends, time in nature) is not essential.**

The more lonely and dissatisfied we are, the more influence the algorithms have over us. This is the game: keep us both engaged and dissatisfied, and continually seeking satisfaction. The platforms want to make sure we get hits of pleasure and enjoyment, but nothing that actually satisfies, nothing that leads us to, god forbid, step away from the platform and live our life. The primary goal is to keep us always seeking, always clicking, always scrolling, day-after-day, until we reach the end of our lives, phone in hand, posting, liking, and commenting as we die. To the algorithms, this is the dream!

**Eric Schimidt, former CEO of Google in a podcast with Scott Galloway, put it quite clearly: "The industry is optimized to maximize your attention, and monetize it."**

That is it. What do the algorithms want? Only your life, or as many moments of your life they can capture. Us going for a run, hanging out with a friend, or spending time in nature is a loss to the algorithm. It is not that the companies do not care at all about our well-being, it's that they care much more about monetizing our attention. As a result, our attention becomes controlled and dictated by trillion-dollar corporations whose primary aim is to monetize it. Don't believe me, believe them!

However, we cannot blame social media or the algorithms completely; we make our own decisions. There was war, violence, and addiction long before we had phones and social media. And there is also a beauty to this situation, as it invites —perhaps even demands—that we connect with what actually matters in our life. This is the only way for us to engage wisely.

When we don't know who we are, we look to the world to fill us. We believe there is something called happiness that the world is holding back from us, and if only we could obtain the right things—more likes, more partners, more status, more money—we would finally get what we deserve.

We think, "If I can get *this title*, *this number* of followers, *this money* in the bank, *this amount of popularity*, then ... I will at last have a feeling of success." Of course, it is fine to get all the above. Yet what determines whether we are satisfied with all this or miserable with all this? Our inner life.

**The solution is simple and likely the most difficult of all endeavors: to know who you are and what truly matters ... and be guided by that. We can then play the game instead of getting played by it.**

When we change, the world changes. When we open, the world responds.

"When the flower blooms, the bees come uninvited."
— Ramakrishna

# WINNERS & LOSERS

"I know, we will just dumpster dive our way across Japan," my friend Taz said, as a group of us sat with our backpacks at a park outside of Tokyo, Japan.

We were on the last leg of a three-and-a-half year global environmental walk. About three years previously the group had begun walking in Los Angeles, and we had arrived in Tokyo to venture about 500 miles to our final destination, Hiroshima. I was on the walk for about a year in total, walking through parts of the US and Asia. The initial plan in Japan was to walk with a group of Buddhist monks and their supporters, who would arrange food, housing, and take care of our needs.

After about a week, however, the monks told us we needed to leave their walk. We had been too loud, failed to walk in line with the others, and generally disrupted their group's flow. As a result, suddenly we were on our own in Japan, with little money, no support network, no housing, and several months of walking ahead.

Even under the best conditions—with food readily available, shelter arranged, and support vehicles in place—walking across a country is a formidable challenge. Strip all that away, and it becomes a true test of survival dealing with changing weather conditions, blisters, and much more. In Japan, each day we searched for shelter, often sleeping in parks on cardboard and improvising ways to eat with what little money we had. At the time, Japan was expensive—a single apple cost several dollars, and a couple of nights in a modest hotel would have drained us completely.

Given our limited resources, Taz was right: dumpster diving was our best option. But on the "status scale," eating food out of dumpsters ranked very low, at least in the middle-class world I grew up in. It is one thing to be poor, quite another to actually reach into a dumpster for food that someone else has discarded.

I had always wanted to be "somebody," to win at the game of life, and this seemed like a huge step backward. I did not disagree ethically with eating food that others threw away. From an ecological perspective, it made a lot of sense: it saved food, reduced packaging, shipping, etc. What made it painful was my self-image. Is this what had become of me? What would people think? What would my family think? In high school, teachers never quite expressed it, but the message was clear: whatever you do, at least have a job, and never get in a situation where you need to eat from the garbage.

Either way, we had no choice. We got up from where we sat in the park and kept walking.

The first week was rough. One night, we dug old McDonald's hamburgers and French fries out of a dumpster. Another night, we gathered leftover potatoes from a field and found a restaurant kind enough to let us cook them. At times, it was pretty gross. However, over time we got better at it. We discovered that convenience stores threw away wrapped prepared sushi dinners every night, and for the next two months, about

90% of our food came from this source. Though we had become accustomed to it, I'm guessing most people looked at our sweaty, dirty group of walkers digging in trash for food and thought, "What a bunch of losers!"

Were we losers? Clearly, most people who saw us digging through the trash would have thought so. But was that who we were? Or were we environmental activists who were actually "winners," because we were trying to raise awareness about the planet, and had such dedication to our cause that we would sleep in parks and eat from dumpsters? What label best fit us? And who should decide what that label was?

It is easy to focus on labels, on who is winning and who is losing, on how things appear to others, but fortunately there is a deeper place from which we can live.

A person I know started giving talks and leading workshops on spirituality in his thirties, and in the beginning only one or two people showed up at his events. But he led these events with as much heart as he would if the room had been packed. He knew that the number of people in the audience did not determine his worth or value. And he was certain that he was on the right path.

Over time, his books became international bestsellers, and eventually thousands of people showed up for his talks and workshops. He packed auditoriums across the world. However, he also knew that this larger number of attendees said nothing about his worth or value. It did not make him any better or more important to have thousands of people in a room than it did to speak to one or two. He had learned to tap into a deeper source, independent of how many people were present. He knew what truly mattered when one person was in the room, and he knew what truly mattered when thousands of people were in the room. The outer shifted, but the inner compass stayed the same.

Sometimes the challenges that come our way are having only one person in the audience or one customer in a business, or losing significant amounts of money, or getting fired from our job, and people see us in a negative light. And other times we may be doing the very same actions, but conditions have changed and more people know of us or buy our product, and our business makes a lot of money. People then think, "Wow, you are winning."

**The real issue isn't that other people put labels on us, it's that we believe those labels.**

It is easy to get lost in the stories society has given us or that we have taken on. When we lose touch with ourselves, we believe, "If I move to a smaller house, it means I become smaller; if I lose followers on social media, it means I am a loser; if I make less money, it means I am also less."

This is the mind that has lost access to its own great nature, caught in the stories of the world, and out of touch with the inherent abundance of life. We become distant from what really matters, from who we really are, from what is essential. We can never satisfy the ever-changing perceptions and stories of the world.

And this is the point: the goal is not to be perfect; it is to find that within us that is perfection, and to live from that place. This is the path.

We can never *become* enough. Time is not involved.

We can only wake up to who we really are.

# THE CURRICULUM

W e all likely have a similar story: A life that has had its share of connection, love, and joy, and also moments of heartache, loss, and pain. Some of us have had more joy or pain than others, and some with a level of trauma that is almost unimaginable. This is what we all have in common, rich and poor, famous and unknown, healthy or sick: we all suffer at times.

Focusing on certain experiences more than others, we can tell variants of our life story. There is the story of how much love we received, and the story of how much love we missed; the story of how successful we are, and the story of how unsuccessful we are; the story of how happy we are, and the story of how unhappy we are. And all of these have truth to them.

However, even when life starts out pleasant and easy, at some point we realize that no matter how hard we try, there is no escaping a certain degree of pain, hurt, and trauma. We cannot control what comes our way in life, and at times what comes our way is very, very hard. No one gets a free pass through. Pain and suffering come.

When the hurt first comes, we do not always have the capacity to be with it—to feel what needs to be felt, and express what needs to be expressed. It is beyond our ability to be with the pain, especially as children, so the hurt gets buried. It stays deep inside, and directs our life, causing us to react and isolate, to lose our temper or run away.

Often our identities form in response to these experiences. We become strong to cover a feeling of weakness, seek status to hide a sense of unworthiness, be "nice" to cover rage, gain knowledge to disguise ignorance, or try to be successful to cover feeling unimportant. The identity we develop often depends on how the hurt came to us.

**Once we experience hurt, often our strategy is to protect ourselves from experiencing that same hurt again. This can be the driver of our life.**

We harden and protect, exiling certain parts of ourselves in an attempt to shield us from future pain. We convince ourselves that if we can just protect more effectively the next time, we won't have to face that same hurt again. But the challenge is that in locking away our pain, we also hold back our joy, love, compassion, and sense of purpose. When we lose access to our grief and pain, we lose access to our joy as well. And without that connection, it becomes incredibly difficult to contact or follow what truly matters.

Our work is not to view ourselves as "someone who was hurt" or to cling to long-lasting resentment toward those we perceive as having hurt us. Instead, it is to gently uncover and understand what prevents our heart from fully opening, what limits the free expression of our true nature, what holds us back from living the life we came here to live.

We each have a path, and the clues of that path are in part discovered in our life events, both the pleasant and the unpleasant, the expected and the unexpected, the welcomed and the unwelcomed. If we are looking for a great teacher, here it is, right in front of us, in this very life of ours.

This is the curriculum.

Our journey usually starts with pain, loss, and heartache. No one is reading this book because life went exactly as they had planned.

We often try various approaches to remedy our suffering, hoping that taking substances, achieving a particular status, or harboring a silent resentment toward life might dissipate it.

Then at some point we realize:

No change in our external world can ever satiate what is unresolved in our internal world.

Then the path opens.

# PART II

## THE POWER OF STORIES

At first we believe that we are our thoughts. Then we see that we are also energy, and that our past and our future spring from the same place, the present moment.

We initially need to see what we are carrying, what stories or narratives may be running our lives.

We do this not to criticize or judge ourselves, but to gently explore how we can begin to liberate and open our hearts.

# THE OPENING

I didn't really know what to do. I was 12 years old, and my family was going through a time of turmoil. For most of my life, we had been a loving, close, and strong family, but things were changing quickly. My parents were spending less time together, and the tension in the house had been growing.

One evening, all five of us kids were called into our tiny living room for an important announcement. Sitting together in a circle on the floor, we were told that our parents were separating. My dad would stay in the house with us, while my mom would live elsewhere. I couldn't fully grasp what was happening, but an overwhelming sense of loss, sadness, and rage swept over me.

While it was personally painful and confusing, it was especially hard for me to know how to explain this change to others, as we were one of the first families in our community to experience a divorce. When kids came over to our house and asked, "What is happening with your family?" or "Where is your mom?" I did not know what to say. I hated those questions.

To deter those questions, I isolated myself. The fewer friends I had, the fewer people who might see the pain that I was in or ask me how I was. At a time when I could have benefited from community, I pushed it away.

I was not able to process what was happening, and isolation seemed like the best approach. I did not want anyone to see the pain I was in because *I* did not want to see the pain I was in. Sleep was one of the hardest parts of the day. My mind would race at night as I tried to process emotions I had never known before: sadness, rage, grief, humiliation, shame. I felt a mental storm of feelings bombarding me incessantly, with no way to stop it. The world I knew before was gone, and it sucked.

My intention and focus at this time was clear: stuff everything down. My heart hurt in a way I never knew possible, and I just could not go on with life as I had before—living as if my grades mattered or success in sports was important. Who cared if my team won a basketball game when I had so much pain in my heart? I was at a loss for how to address the depth of hurt I felt. No one had told me this level of suffering was a part of life. I was pissed off at myself, my parents, and most of all … at life. How could life that I had so trusted let this happen?

Then one day my father left some books and tapes on meditation, spirituality, and Buddhism outside my bedroom door. Since I shared a room with my brother and sister, I would retreat to my father's bathroom, close the door, and listen to guided meditations. It was the one strategy that seemed to help ease the constant chatter in my mind. After many months of practicing, one day during a meditation I experienced a moment, though very fleeting, of non-chatter, of stillness, peace, spaciousness.

My mind, for a moment, was at ease. Within the flood of emotions and feelings I saw that there was a deeper dimension —a dimension where there was no problem for me to solve, where thoughts and feelings arose and passed away in a type of calm, spaciousness. There was not only chatter, there was also space.

I realized that humans had a power I never knew was possible: the power to be with whatever is emerging in a way that does not add more difficulty and pain. The pain is there, but something else is also there, a presence or spaciousness. These were the two significant realities that started to hit me: intense pain, suffering, and trauma are a part of life, AND there is some part of us that cannot be injured or destroyed. Call it what you wish, but there is also a spiritual dimension. Suffering and freedom from suffering both exist.

**To be human is to dance in these two worlds.**

I once met a man in Rwanda who lost all 11 of his siblings in the Rwanda genocide over a few months. I have also been with kids in juvenile hall who, at 12 or 13 years old, had watched friends get shot and killed. This is our world. Every day, thousands of people across the globe are killed, brutalized, or die of starvation. There is such pain and suffering, from lack of safety to starvation. At the same time, a wealthy child in America who has everything materially, but is not receiving love or attention from his parents, also hurts. He or she feels the hurt of neglect, and that is just as real as any of the other hurts.

The issue is not that hurt or pain comes our way, it is that we often close around the experience. We close our hearts to ourselves and to life. And this in many ways starts our path. We then have the great blessing of opening and connecting again to who we are and what matters.

**When we reflect, we often realize that closing our hearts was the most compassionate act we could have done.**

We can do our best to live in a world of incredible suffering, and to meet that suffering with as open a heart as we can. The spiritual path to me is equal parts pain and joy, and requires a willingness to see what we would rather not see, to go where we would rather not go, to feel what we may not want to feel … and *at the same time* to know the essential beauty that we each are.

To better know who we are, it helps to see who we are not.

This process can be uncomfortable, as old patterns often cling to familiar ways.

While one part of us resists, if we listen more deeply, another part often responds, "At last."

# THE POWER OF STORIES

"**D**o you actually believe that you are unworthy?" my therapist asked me. I had recently been through a divorce and I was trying to better understand why my intimate relationships with women were so painful. I had reached out to a therapist for help.

I wanted to say no to her question, but deep down, I knew the real answer was yes. I had accomplished a great deal in life by that time. I had achieved many of my goals, earned praise, accolades, awards, and much more. I had reached a certain level of social status. Yet, the truth was that, despite all the outward success, all the praise I had received, all the people who respected me, some part of me still felt unworthy.

As my therapist and I delved into the question more, it became clear: a part of me remained in that painful time as a child. When my parents separated, I came up with a story: "The reason they broke up was because of a flaw in me." Until she asked me, I had never thought about it. But at that moment, the question hit me deeply, and I could not lie to myself or deny it any longer. Though I had achieved so much, there was still a part of me that felt inadequate, unworthy, and unlovable... and this part was directing aspects of my life. The external praise I had received did not heal the internal feeling of unworthiness. In fact, it merely covered it.

This story of unworthiness was, of course, wrong, but also quite reasonable for a child of 12. That is what often children do; they take events personally. And though I was now an adult, the story had lodged inside of me, and had directed my intimate relationships. While I was an adult by age, some part of me was still that young child. I had, in many ways, forgotten and abandoned him.

To hide the shame and unworthiness I felt inside, I had developed a new story on top of this, "I am a success." This was the story I wanted to present to the world, one in which I was a "winner," popular, respected, hard working, and someone who had made it. Sure, some of my work was infused with care and love, but there was an element of wanting to prove that I was in fact worthy. If I had enough status and success, and I knew a lot of important people, maybe this unworthy part of me would be healed ... or so I thought. It seemed like a good strategy at the time!

I began to understand the power of stories.

As we look within, one of the insights we often have is how much of our lives and our world is impacted by our thoughts and the stories those thoughts construct. Just stopping for a minute and sitting silently with ourselves, we see what the mind does: it creates stories—about everything! The stories cover almost every area of our life, and usually get expressed as: "I am this," and "I am not that," or "I cannot be this," and "I must become that."

These are the stories of "me."

In fact, every religion, every political system, every economic system is a story. Someone has a thought, or a series of thoughts, then convinces enough people to believe those thoughts, and all kinds of systems and religions come into existence. All of a sudden we have Christianity and Buddhism, capitalism and socialism, liberals and conservatives, and so much more.

Once these stories are integrated into our societies, we tend to think, "This is the way it is." But it is only the way it is because people believe it to be that way. These are stories people have come to take as truth. It does not make these stories any less important or useful, but it helps to understand that these are all mind created. Essentially, the world runs on stories.

**It turns out, the story we have of who we are is not actually who we are.**

Some of the most powerful and impactful stories are the ones we have created about ourselves. For example, we may think, "I am wealthy" or "I am poor." Or "I am smart" or "I am not smart." But are these actually true? Can any of us prove we are wealthy, poor, smart, or not smart? Someone might say, "Yes, I am poor as I only have $10 in my savings," or "I am wealthy because I have one million dollars in my savings." But are these statements actually true?

For example, someone might have $10 in savings and believe the story, "I am a success." She believes this because she grew up with enormous debt, and the fact she is not in debt makes her feel like a winner. Even though she only has $10 dollars to her name, she walks around thinking, "I am a success." She no longer has debt, and for that she thinks she has made it. That is her story.

Someone else who comes from a very wealthy family might have a million dollars and walk around thinking, "I am a failure." Maybe he lost $10 million and only has one million left.

Or his family has tens of millions and they view him as a loser. He was expected to add to the family money, not lessen it. Furthermore, he primarily spends time with people who have much more money than he does, and who can purchase better clothes, fancier cars and bigger houses. Because of all this, he believes, "I am poor."

You may counter, "That is delusional. If you have $10 dollars you are poor, not rich. If you have a million dollars, you are rich, not poor." It *is* delusional to think you have a million dollars in the bank when you only have $10 or vice versa. That is not seeing life as it is. But the story we create about that money is our own and it often governs our behavior.

You may disagree with a story someone has, but who are you to say how someone should view themselves? People make up whatever story suits them, often not realizing that it is a story. Someone has $10 dollars and creates a story of success, and another has a million or a billion dollars and creates a story of failure. We make up what we make up. I know people with billions who mainly live from a place of lack, and other people with almost nothing that live with a sense of abundance. People simply hold different stories.

Of course, we need stories and norms to function in any society. If I believe a red light means to stop and you believe it means to go, chaos will ensue. Our monetary system is based on a story. It works because enough people believe it. Both the buyer and the seller need to believe the same story. If I believe this dollar can buy a banana, but the seller of the banana does not believe that, we have no means of exchange.

Stories can serve a purpose. And on a functional or relative level, we are many things and need to be them: a student or parent or manager or whatever. These are our roles, and we follow the rules of society, understanding there is a functional truth to them. At the same time, we also know who we are independent of these roles.

Now, you might be thinking, "Why does this matter?" It matters because our stories impact our world, and some of the greatest politicians, activists, and business leaders know this secret. And they use stories to both hurt and to help, to manipulate and to inspire. Control the story, and you control people's attention and worldview. The answer is not to try to obliterate these stories, but to bring awareness to them.

**We need stories for practical purposes, but we need not be run by them.**

The stories of who we are, or who we think we are, however, never tell the full truth. Our uninvestigated stories often serve to keep us locked in our thoughts, and cause us to miss what is most alive and true for us.

As we wake up, we begin to inquire, "What stories do I hold about who I am and how the world is? And are they true? Do they serve to expand or limit me?"

In this inquiry, a door opens.

# THE CONVINCE

I realized that once I took on a particular story about who I was, a "story of me," I had to convince other people of this story. This is true of any story. It is not enough that we see ourselves as wise, smart, loving, beautiful, successful, powerful, funny, tough, sexy, unhappy, wealthy, and so on. We need to convince other people of this.

Every story needs to be fed. And it is fed by other people believing that story. This is the story's fuel source.

There are ways to engage and share, whether in person or online, that are truly authentic. We are simply being ourselves, with no special game at play. We value having a voice and use it to share with friends and others what we are doing and what truly matters to us—and that is beautiful.

Yet there are also ways to share that attempt to reinforce a story. What story? Any story! The story that we are smart, the story that we are spiritual, the story that we are successful, or that we are thoughtful … you name it. There is an image we want to put out. This is why people who identify as "spiritual people" can be so freaking annoying sometimes. They only see themselves through a certain light.

This plays out on almost every social media platform: we have an image, a story of who we think we are, and a desire to have that image reinforced and believed by others. It may not be the whole reason we are on the platform, but it often is a part of it. And the more likes we get, the more we keep going. Our image is continually fed, but never satisfied.

**Life can be consumed with one question: How do I get other people to believe the story I have about myself?**

Look at social media from this perspective and it makes a lot of sense. The goal is to reinforce "the story of me" out in the world, hoping others will believe it. Everything we share (or don't share) has to go through this test: Does it reinforce the image I have of myself? If so, we share it; if not, we rule this story out. Thus we curate an imagined life.

Out of 1,000 moments in a day, we get to pick the one moment that makes us look good and post that. All the other parts of our lives can stay nicely hidden from view. The argument with our husband or wife, the yelling at our child, feelings of pain and misery at times all can be easily glossed over because, "Look! Here is a picture of all of us smiling together! Aren't I amazing?"

Social media can be much more than this, of course. It can be a way of sharing truth and depth, and updating people on meaningful life events. But the tendency, if we are not careful, is for it to become a status game, seeking likes and approval without touching what actually matters.

**Freedom is the experience, not the thought, of who we are.**

However, as soon as we see the pattern, we are free. We see that deeper than our story lies our essence, our true nature, our full expression. And this does not fit neatly in any box or role. In fact, the more we try to control or manipulate our image, the more we try to convince others to only see us in a certain light, the further we get away from our essence.

Who we are is not an image or a thought. We must go deeper.

# IS IT THE THING OR THE THOUGHT
# THAT MATTERS?

Our world is filled with ways to avoid and escape, and our phones provide endless options. Advertising tells us, "Feeling unhappy? Just buy this, eat that, or watch this." The result is that our feelings never get the attention they need. On the other hand, we try to amplify pleasant thoughts and experiences, seeking activities that give us a dopamine boost. We both push away what we do not like and try to grab what we do like. This pattern of pushing and grabbing, it turns out, is exhausting.

Much of what we are resisting or clinging to is ultimately tied to the thoughts we hold about ourselves and the world.

There are billionaires, for example, who own mega yachts, sometimes more than one. These yachts can cost hundreds of millions of dollars, take years to build, and require enormous resources to maintain. Sometimes the owner only uses the yacht a few weeks out of the year. Why would anyone want to own something they use so rarely? It's not about the yacht. Sure, the owner may have a good time on it, but they could easily rent one. The owning of the yacht serves another purpose.

What they get out of owning a yacht is they can now believe the thought, "I own a yacht! In fact, I own a massive yacht, bitches." Even if they don't say it aloud, they know it and feel it. The thought is what matters most. It serves them in various ways. They can casually bring it up in conversation: "Well, you know, I have this yacht," or "When we go to our yacht in the Mediterranean, it is so much fun." This reinforces their sense of importance and specialness.

This makes them feel like a winner, someone of importance, a person at the top of society. To feel this way, they don't actually need to use the yacht. That's not the point. What really matters is that they can think of themselves in a certain way. The ownership of the yacht is simply permission to believe a thought. That is all. They want to hold a particular story of themselves, and the yacht gives them permission to do that.

**Possessions or accomplishments can at times be a way for people to feel permission to believe a thought.**

Now, there are certainly some people who actually enjoy their yachts, and it is not about the identity. But the tendency, especially among people of wealth, is to create a better story than the story other people hold. And yachts, third and fourth homes, country club memberships and certain watches are a popular strategy for doing so. Possessions themselves aren't inherently problematic; the problem arises when we depend on them to fill an inner void, to add to the story of me.

And while it is easy to think this is simply an issue of the wealthy class, we can all inquire, "How do I use what I own to reinforce a particular story? And does this serve me or not?" It is yachts in some communities, cars in other communities, hair styles in some, and scars in others.

How much of our life is focused on trying to create conditions that give us permission to believe certain thoughts?

When we do not need to reinforce these stories, and use possessions to help us feel better about ourselves, we can tap into a deeper intelligence. When we are connected to this, we have nothing to prove and no one to impress.

This is freedom, the rest is a continual race to nowhere.

# THE HIDING

As a child, I never learned how to ride a bike. That skill somehow escaped me. There were five of us kids. We played a lot of sports, and finding time for someone to teach me how to ride a bike never happened. Once I had missed the window of when the average kid learned to ride a bike, say at 8 or 9, I knew of no other strategy than to try to hide this fact from everyone. I wanted to be seen as cool, and not knowing how to ride a bike, I worried, would reveal that I was inadequate.

Of course, at any time in my teen years, I could have said to a friend, "You know, I do not know how to ride a bike; could you teach me?" Likely, I would have learned quite easily. Problem solved. But in those days, I hid the shame I held inside for not knowing how to ride, that no one had taught me such a simple skill everyone around me had mastered years earlier.

**Shame leads us to hide, yet slowly we can learn to open again.**

This same shame also showed itself in my sexuality. Many of my friends were experimenting with intimacy and sex in their teens. The majority of my friends had had sex by 16 or 17, or they had at least kissed someone or gone on a date. I had done none of this. As I passed that age, and the topic came up of how many girlfriends or sexual encounters people had had, I became very skilled at changing the topic, leaving the room, or avoiding the conversation. I did not have the capacity to name the truth: "I have not had sex or much of any intimacy with anyone."

This inexperience continued into my twenties. "Hey, Soren, so when did you first have sex?" my friend asked me, as we were sitting around the room with four or five other friends. I was about 21 years old, and the honest answer was that I had never had sex. I had kissed maybe one or two women in my life. When I heard the question, one part of me thought, "Just speak the truth. Who cares?" And another part screamed in my head, "Avoid, retreat, lie, get away, danger! You don't want people to see that you are inexperienced; imagine what they would think."

My more honest answer, and the one I kept hidden, was that I was afraid of women, and the thought of surrendering to someone in that way was terrifying. But it was too hard to acknowledge this truth to myself, much less to someone else. I felt so much shame, and I did not want people to see that. I wanted to be seen as cool and savvy, and my lack of sexual experiences did not fit that image.

My way of coping with uncomfortable truths and situations was to hide. Sometimes this meant canceling outings or avoiding friendship with anyone who asked direct questions. Anyone who spoke the truth terrified me; I worried they would force me to do the same. It was easier to avoid and hide.

At some point in my mid-twenties, a friend asked me to go bike riding with him, and when I told him I did not know how to ride a bike, he responded, "Amazing, want to learn?" We went to a local park, he gave me some lessons, and after about two hours, I could ride a bike. Ten years of shame for something that took about two hours to learn!

While there are certainly situations where it is best to withhold and keep things private, the challenge occurs when this pattern becomes automatic and enduring. Even when conditions are conducive and safe, we decide to hide. We keep parts of ourselves exiled from others, but most importantly from ourselves. Hiding becomes a habit and potentially an addiction.

It is just how the pattern works. "Look at me this way," inherently means, "Do not look at me this other way." Or "Please see me as confident and capable," means "Please do not see me as insecure and uncertain." For me, the parts I didn't want others to see were my unworthiness, weakness, vulnerability, and helplessness.

These are often what my friend Dick Schwartz calls "exiled parts," aspects of ourselves that we have pushed away and neglected. Often a past painful experience made us stash these parts away, making them temporarily inaccessible to us. This happens automatically to most of us. We need not berate ourselves for this, but to connect, to learn, to forgive, and to find ways of opening so our delicate spirit has more space to shine.

**When parts that were once hidden in the darkness are allowed into the light, we now have access to them.**

There are even spiritual teachers who appear profoundly wise (and often are in many ways) yet have exiled parts that eventually surface, leading to unskillful actions such as stealing money, misusing power, or engaging in sexual improprieties. These repressed parts did not align with the image these leaders sought to present. Such patterns can remain dormant, only to emerge suddenly during moments of stress, challenge, or after achieving a certain level of status or success. They were not gone, just dormant.

How do we see the patterns? We pay attention. And we find the practices that best support us, be it meditation, therapy, Internal Family Systems (IFS), psychedelics, or something else. But waking up is not confined to any system or tradition, and our heart usually knows the way.

No matter what we do or do not do, eventually whatever we hide comes into the light. The more we accept this and recognize that we are all of it, we are on the path.

# THE PLAY OF PARTS

O ne of the most challenging realizations I've had is recognizing how tenaciously a part of me clings to my story, even when it undermines my well-being. In my case, my desire to be seen as a success meant I would subtly (or not so subtly) attempt to affirm it at the expense of others. I would compete with others for status, feel resentment toward those who had achieved more than me, and go toward those people who I thought could help me be a success and ignore anyone I thought could not.

We all have our own patterns, of course. If we view ourselves as "the responsible one," for example, we tend to judge those who are less so. When people fail to follow through or take care of certain responsibilities, some part of us may be thrilled. "I cannot believe there are all these irresponsible people in the world," we might think. And the truth is while some part of us might be frustrated by other people's actions, another part is quite happy.

How could we be happy about someone's lack of responsibility? Because that person's lack of responsibility reinforces our view of the world and of ourselves. "I am the responsible one, and others are not living up to my standards. I take care of things, and other people do not." We love it when life reinforces the view we carry of ourselves. It gives us a role to play. In fact, we may look for actions we view as lacking responsibility so some part of us can feel righteous.

As I mentioned earlier, all identities need to be fed, they need a fuel source. In order for us to feel "more responsible," we need to have people around us who make mistakes. Otherwise, our identity as "the responsible one" is very hard to sustain. How can we see ourselves as "more responsible" if others are acting equally responsible?

To feed our "responsible" identity, we may tell our friends, "You would not believe what I had to deal with. My coworker was supposed to do this task, and he completely blew it off. And so yet again I had to come to the rescue and work extra to finish it. I had to save the day." Or "My child just will not do his homework, so I had to sit with him last night. I am always the one who takes care of things like that."

Within this pattern, we focus on other people's actions, and if we make a mistake, it is hard for us to see it. Our view of ourselves does not allow us to see any of our failings. When we do make a mistake, instead of thinking, "Wow, I make mistakes too. I see how that is what humans do sometimes," we gloss over anything we've done wrong so that we can keep our identity intact. After all, how can we feel better and more responsible than others if we are all human and all have flaws? So we divert our attention from ourselves, and focus on others' mistakes.

This dynamic manifests in many ways. For instance, if we see ourselves as a helper or a hero, we often unconsciously need people around us to "save." Often, we subtly set up conditions where a dependency forms. Eventually, we can start to complain, saying, "Stop needing me to save you," to someone who regularly seeks our help. Yet, another part loves it and creates conditions for that person to be dependent on us. So while one part of us resents their dependence, another part craves it. It's a paradox—complex, relatable, and surprisingly common.

If we cannot see or acknowledge this other part, it works underneath our awareness. For me, this has been one of the hardest aspects of myself to see. "No, I do not want people dependent on me," I tend to say; yet if I am honest, I create conditions that encourage that dependency. I become dependent on them depending on me.

If I am not aware of this, I will continue to blame everyone else, never seeing the part I play. The pattern persists. And if one dependent person leaves my life, it will not be long before I find a new person to play that role. People leave, but the roles they play are soon filled.

I often ask myself, "Is there a part of me that actually enjoys what I complain about?"

Usually there is something for me to understand more deeply. There is a part of me that needs my attention and care. And in that compassionate seeing, our capacity increases.

# THE TRUTH

"You just don't get it, dude!" the young man yelled at the Dalai Lama, the spiritual leader of Tibetan Buddhism. "The system is set up for us to fail. They make more money on us in prison than out of prison."

I was sitting in a swanky midtown hotel in New York City with about 25 former prisoners, a handful of Tibetan monks and Westerner teachers, a number of U.S. State Department Diplomatic Security Service agents … and of course, the Dalai Lama.

"I was in prison so long, when I was out, I would stand at the street corner, waiting for someone to tell me it was OK to cross, because that was my life inside," the young man continued at a pace few of us could follow. His words had fire to them, and the tension in the room continued to rise. As his anger intensified, the State Department and Tibetan security details seemed to be particularly alert.

I wondered if the organizer of the event would step in. Then I remembered, *I* am the organizer! And in many ways, the young man was doing what I expected.

Almost everyone in the room knew of and had been impacted by the Dalai Lama's teachings. But in organizing this session, I wanted to include a few people who came from other communities who might not be as familiar with him or his stature, and who might speak more bluntly. I had found a young man (the one who was yelling) named Xavier who I thought would speak his mind no matter what.

When I first met Xavier in the Bronx through a contact at a juvenile justice non-profit, and asked if he knew of the Dalai Lama, he said, "No, who is he? Is he like the Pope?"

I explained who the Dalai Lama was and invited Xavier to join us for a meeting. While most people would jump at the chance to meet the Dalai Lama in such an intimate setting, Xavier couldn't have cared less. He simply shrugged and said he couldn't make it because he had a meeting with his parole officer at that time

I told him that I would talk to his parole officer and get permission.

At the event, the Dalai Lama had asked to hear about the lives and experiences of those in the room, and Xavier had responded, honestly and directly. And the Dalai Lama listened. The Dalai Lama spoke the following day about how this conversation had impacted him, and allowed him to gain a deeper understanding of the challenges within this system. This occurred because Xavier was focused on the truth more than on being nice. He prioritized what was real instead of what might be pleasant.

**The truth is not always easy, but it is our starting place.**

I learned this the hard way.

"Have you ever asked your son what his experience was?" My therapist questioned me at a difficult time in my life. I was going through a divorce, and had disregarded my son leading up to it. I had ignored him and I was not there for him when he needed me. As a result, there was still a lot of tension between him and me. I wanted everything to be better and convinced myself that if I just pretended everything was fine, surely it would be.

"I have apologized to him numerous times. I have told him that I am sorry," I told her.

"Apologized for what?" she asked. "Do you know how your actions impacted him? If you have not taken the time to ask him about his experience and really heard his response, what are you apologizing for? He needs to know you are willing to hear directly from him what this was like for him. Otherwise, your apology means nothing."

She had a clear point: there is no way to heal or get past something we have first not fully seen, heard, and accepted.

I went to my son and asked him, "How was the divorce for you? I do not think I ever asked you about your experience, and I would like to hear it. I am curious how it also might impact how you see intimate relationships."

It was so hard to be curious about his experience, and a part of me really did not want to hear his answer. But once I actually asked, I felt tremendously relieved. As I was able to hear his experience, even the parts I did not agree with, the tension between us started to decrease.

This tendency has shown itself again and again in my life: I want to move on before being with and feeling what is. The healing, I learned, is in acknowledging what is true and feeling all there is to feel, not trying to move past it.

**Buddhist scholar Bikkhu Bodhi said, "We have to be prepared and willing to discover what is true, even at the cost of our comfort. For real security always lies on the side of truth, not on the side of comfort."**

The whole spiritual path could be broken down to this: What are the truths of life? And how do we live aligned with these truths?

# THE FAMILY AND CULTURAL STORY

○

We each have personal stories and identities, but so do groups of all kinds, including families, clubs, countries, and virtually any collective, in fact.

One day while teaching my weekly class at a juvenile hall in New York City, I passed around a bell for the kids to ring. It was a beautiful bell, with a small wooden stick for striking it softly to invite a sound. The kids loved the sound and it gave them an easy way to practice focusing their attention. As the bell made its way around the circle this particular day, one kid grabbed it and hit the bell really hard with the stick. I heard a thud; the bell had cracked. I was visibly sad when I heard the thud and saw the crack; I loved that bell. Then, as soon as the kids noticed my sadness, they came at me. "Oh, you sad, Mr. G.?" (they often referred to me as "Mr. G".)

"Nope, I'm not sad," I replied, lying to them. Of course, I *was* sad and they could see it clearly, but I fell back on an old, well-worn strategy: denial. They came at me more. "You need your mom? Oh, you poor baby. Look, his bell broke and he is sad. Do you need to cry?" Numerous kids ganged up on me, all laughing and telling jokes at my expense, pointing at me. They could see I was hurting, and in response, they attacked.

At first, I was frustrated and angry, thinking, "Really, I am sad because my bell broke, and you assholes make fun of me! That is cruel. Who would do that? I started a non-profit group to help you, and I have volunteered hours and hours of my time, and this is how I am treated? I am such a good guy, and I am treated so harshly." I had to let my initial reaction move through me.

But I knew these were not inherently mean kids; something else was going on. It then became clear: it was not just the bell that had broken. I had broken one of the codes in juvenile hall: never show sadness or weakness. This was one of the foundations of that culture. They could not allow me to be sad because they could not allow themselves to be sad. The story in the facilities was, "Be tough, no matter what. Never show vulnerable feelings."

**Most groups have a story that guides them. The story is often hidden until we intentionally or unintentionally come up against it.**

My sadness challenged their narrative about toughness, leaving them little choice but to respond the only way they knew. Though it was hard for me at the time, it was a powerful lesson: when someone expresses something that contradicts a group's narrative, the group feels compelled to defend its shared story.

What the kids were actually saying is, "The experience you are revealing is not allowed in our mental model. We do not know how to keep our sense of self AND incorporate sadness. For our own protection, we must resist and make fun of you. It is not personal, it is how we need to survive."

Groups tend to move toward what reinforces the story of that group, and move away from what questions or threatens that story.

Either way, be it going towards or pushing away, there is usually a particular charge to the response. We can feel the intensity. It is either, "Yes," and there is an attraction toward or a "No," and there is an equally intense force to "resist." Both are clues that something else is present in the room. A story is at play.

In the San Francisco Bay Area where I live, money and social capital are certainly important, but the strongest motivation is to be an innovator. We are constantly trying to create "the next big thing." Money we inherit or simply keep in the bank is often seen as unimportant or weak. Who cares? What are you doing to change the world with that money? How willing are you to risk your money? What great idea are you backing to change the world? These questions matter more than money.

We see ourselves at the forefront of creativity and innovation. After all, this is where Apple launched, as did Meta, Google, and OpenAI. The competition is over innovation and a willingness to risk. Our story might best be summarized as, "We are the innovators. We are the most creative people changing the world."

The story drives us. But is it true? Of course not. No one person or group is any single thing. Is a more ordinary job, like a plumber or deli worker, less important? Absolutely not. There is nothing inherently better about being a founder or entrepreneur than being the person who decides to live a much simpler life as a gardener or electrician or waiter.

**No role in life is inherently better than any other.**

But we can easily forget this deeper truth. We don't realize, "Oh, this is just a story! I can choose my own path. I don't have to be an attorney, if I prefer to be an artist, or vice versa." Historian Yuval Noah Harari described it as a suitcase that is handed down to us, with all the stories and beliefs of previous generations. Maybe we want to keep some, and there may be some we want to clear out, but if we do not become curious about what is in the suitcase, it will direct our lives.

Families also have stories. The family may have a myth, held silently or expressed: "We are successful," or "We are well-dressed," or "We are anarchists," or intellectuals, or entrepreneurs. The children in the family have to negotiate between what they take from the family and their own natural inclination. They may spend 25 years trying to be successful because the family myth pushed them to be so, only to realize later in life that this was the family story, not theirs. The motivation of past generations may not be ours.

The family myth could also be, "Achieving is bad," or "We are losers. You can never amount to anything," or "Fight the system. Never fit in." The story is to fight and resist the current structures, no matter what they are. Children may believe this for years, then at some point ask themselves, "Is this really true? Are we really losers? Do I really have to fight the system? Or can I choose something else?"

We take on stories from numerous groups, but the question is, "Are these true for me?" If we carry the family story "We must succeed," then we can never rest. If we carry the story, "We cannot succeed," then all we do is rest. These are not bad or wrong, and we cannot blame our parents, the society, or anyone else. The stories simply came to us.

As we grow, we become more curious about the stories that we have inherited from our parents, cultures, and past generations. We can inquire:

**"What are the family and cultural stories I have inherited? And are they true for me?"**

# ME AND WE

W hen we are unaware, everything is about the story of "me." It is not that we intend to make this our focus, it's just the pattern of the unconscious mind. Two minutes into a conversation, instead of deeply listening to another person, we jump in and say, "Well, let me tell you what I think about that."

I find myself doing this often. I shift the conversation back to myself, as it is too hard to not be the center of the conversation. This is simply a product of our conditioning. We are taught and trained to reinforce the story of me, and conversations can serve as a vehicle for this reinforcement.

This can take many forms. A friend might tell us about a particular challenge, saying, "I sprained my ankle really bad and I am in a lot of pain." And instead of listening and inquiring more, our first response is, "Well, I actually broke my leg once, and talk about pain, that was really intense. Let me tell you more about it …" And off we go.

It is hard for us to take someone in, to put ourselves in their position, and respond, "I am so sorry. Tell me more." Again, it's not really our fault. It's just a habit of the mind.

Of course, we can also have a pattern of never telling people what is going on with us, always holding our feelings inside. We can feel unworthy to share our views. There is no real right or wrong in this; the question is, "Are we aware or not?"

When we are present and connected to what matters, we can say, "I have something to share in this conversation." Or to a friend, "I would love to hear from you, but now is not a good time for me. Can we schedule a time later?" We take responsibility for our experience, and notice the tendency to create a story about the situation.

**It is not that we never have the thought, "What about me?" it is that we are not consumed or unconsciously directed by it.**

It is easy to get caught in a "me" frame. No matter what is happening, the question becomes, "Well, how does this benefit or help me?" Or "I am sorry you are hurting, but this is not pleasant for me. The fact that you are sick is really a bummer for me." Or "you being happy is actually making me sad. This is not good news for me."

I went on a two-hour hike with a friend some time ago, and I do not think I spoke the entire time. It was a non-stop monologue about her world. And I can only guess I have done the same at times. When we get lost in our story, there is no awareness of the other, no attunement, no presence to what is happening inside or around us.

This can also play out with our partners and kids. "You having that job, or dressing how you dress, or having the interests that you do does not make me look good," we might think and say to a child or partner. "It does not reinforce the story of me that I want to present to the world. And since you are affiliated with me, this has to change!"

We may use different words, but that is the vibe. It doesn't really matter if the person is truly happy in a particular job or following a meaningful talent they love or dressing a particular way that makes them feel good ... we see everything through the lens of "me and my story."

A parent watching a child play basketball could watch from a place of curiosity and support, or could be thinking, "Your playing doesn't support my story of me. I see myself as a great athlete, and you are not reinforcing this. We have a problem." Of course, the one with the problem is the parent, not the child. The kid of course feels this disappointment from the parent even if it is not expressed. The story is palpable and painful.

This is quite different, of course, than having values, setting clear boundaries, or approaching someone due to unskillful actions you witness. There is a profound difference between addressing someone with true concern and love versus disappointment rooted in his or her failure to reinforce our story.

**When we can notice our story of me and not be motivated by it, we can actually connect with another human being.**

Of course, what we want and need still matters, but we are not driven by an endless need to bring attention to ourselves. We tend to and understand those parts of ourselves so they do not unconsciously motivate us. We are aware of the "me" and make space for the "we," and we get the benefit of actual connection. We dance between listening and speaking with genuine connection as the focus.

I try to keep it really simple:

When someone is talking, can I look them in the eye and listen … and when I am talking, can I speak what is real and true?

Everything else is out of my control.

# PART III
---
# BEING

There are times when we are awake, attuned to the present moment. We are not clinging to anything or resisting anything.

The mystics spoke of this, as have some of the greatest athletes, business leaders, and artists.

There is an inner intelligence.

And it is here right now as you read this.

In fact, it is only ever here now.

# PRACTICE PRESENCE

W hen we get lost in our stories, we miss something essential: the present moment. There may be no more important practice than cultivating presence, having our attention right here, right now, with spaciousness and curiosity.

Many middle-aged people aspire to live longer, often focusing on the physical aspects of their lives. They prioritize health metrics or fitness goals, such as the ability to run three miles or lift 200 pounds at the age of 70. All this is certainly important. I exercise regularly and I know that eating healthy can have a huge impact on our well-being.

At the same time, how many moments during those three-mile runs or weight lifts are we actually present? And how much of that time are we mentally elsewhere, lost in thought, largely oblivious to what we are doing? We are trying so hard to get "there", we miss what is "here."

Walk into any gym, and you'll often see people pushing themselves to the limit, seemingly driven by a fear of aging or decline. While the intention is understandable, the tension and underlying fear might be doing more harm than the good of the exercise. People talk about lifespan—the number of years we live—and healthspan—how healthy we are as we age—but I also like to consider something I call "presence-span," which is how fully we inhabit each moment.

Would you rather live to 100, present only 10% of the time, or live to 75, fully engaged in 80% of your moments? Most would choose the latter, valuing the quality of each moment over sheer quantity of years.

"Most people go through their lives very rarely being present," Dr. David Sinclair once shared with me. A leading expert on longevity, Dr. Sinclair understood that while extending life is significant, living fully in the moments we have is just as crucial, if not more so. He, like many others, began to ask, "What good is living to 100 if we are not even appreciating the moments we are living? Do we want to reach old age angry, scared, and resentful? Is that truly living?"

**Presence, or mindfulness, is the ability to bring our attention to this moment, right here, right now.**

Most of our lives are consumed with thoughts, generally of the past or the future, which means we miss what is most important of all: this very moment, the only moment we have. We might feel this presence most distinctly when we are in nature, playing sports, or making love. In these instances, we feel a connection with what is happening right now. When hiking in the woods, we might become acutely aware of the rustling leaves, the scent of pine, or the rhythm of our footsteps. When playing a sport, we might feel entirely absorbed in the game's flow, every sense heightened.

It's easy to think that outside elements "make us more aware," and there may be some truth to this. We often feel more present in nature than when spending hours staring at a computer screen. However, while these external elements can help, presence ultimately comes from within.

We can harness this awareness in any activity, from walking down the street to talking with a friend. Suddenly, there's a dimension of awareness of what's happening: we catch our friend's eyes during a conversation, deeply hear their words; notice the texture of the bark on the tree we're passing, or taste our food. Contact is made. We are not trying to get anything from someone, or be seen any particular way, or race to the "next moment" … we are simply here, now.

This is the shared superpower of humanity. By tapping into it, we unlock deeper connections with ourselves and others, enhance our creativity, and experience life more fully. It starts with us.

**In presence, there is no agenda, no attempt to control others, no need to prove anything.**

When we are present, we are here, awake, attuned to the moment as it arises, and responding as the moment calls. We realize that the source of awareness and fulfillment is inside us. No matter what new AI we create, what does it matter if we lose contact with the wonders around us everyday?

When we are aware, we are awake to the present moment just as it is. There may be nothing more essential.

What is the most important moment?

The answer is always the same: Now.

And how are we with this moment, now, the only moment we have?

# KNOWING, WISDOM, AND BEING

There is a story about a professor who visited a Zen Master. The professor began by telling the Zen Master everything he knew about Zen, speaking endlessly without pause. The master listened patiently and then asked, "Would you like some tea?"

When the professor nodded, the master began pouring tea into his cup. She kept pouring even after the cup was full, and the tea began to overflow.

Finally, the professor exclaimed, "Stop! The cup is full—it can't take any more."

The Zen Master replied, "Like this cup, your mind is full. To learn Zen, you must first empty your mind, empty your cup."

The professor had knowledge—he knew many facts—but he lacked wisdom. He knew so much about Zen, but he knew so little about *how to Be*. We could say "he knew the truth, but he could not Be the truth," and the latter matters more.

AI, for instance, provides access to endless knowledge, which is valuable, but there is more to life. Being comes through paying attention to our own direct experience. You can repeat what someone else has said and call it wisdom, but that is merely echoing words. True wisdom arises from the well of our own experience, from paying deep attention to how life unfolds both within and around us.

People who have never read a book or listened to a podcast can have profound wisdom. They learn by planting seeds, tending a garden, watching the sunrise, or simply observing life. They do not need to memorize facts or know current events to prove their intelligence.

**One practice is to pause and ask: "Am I only focused on knowledge and information, or am I also making space for wisdom?"**

Most podcasts and news today spread knowledge. Guests often "know" a lot about their topic, sharing facts, opinions, and beliefs. This is of course quite useful. I want a doctor or a mechanic with a lot of knowledge, but knowledge is limited. We might have extensive knowledge about a host of topics, from meditation to wellness, but how does that help us? It only helps us if we put that into practice; it is not so much about knowing it, it is about Being it.

At times, we can use knowledge to one-up each other: "Well, I read that's not true" or "I actually know a lot about this topic." Knowledge is frequently wielded as a weapon to assert importance and domination. While knowledge has its place, the deeper question is: Where is it coming from? Is it serving the moment, or is it simply a means to impress?

**In Being, we are as comfortable with not knowing as we are with knowing.**

Being is fine with just hanging out. It is often expressed through silence as much as through words. Being exists in a realm beneath words. It is "the space" in a conversation that can be sensed, felt, and accessed. When we are connected to this we do not need to say, "I'm very connected to Being right now. You may notice. You are welcome!" No! That is not it. To say it ruins everything!

**In practical matters, it helps to know. In other areas, it helps to not know, to be open to the mystery, to be curious, to rest in Being.**

Likely the most difficult of all endeavors …

When walking, just walk.

When talking, just talk.

When thinking, just think.

When eating, just eat.

When listening, just listen.

When living our moments, just live our moments.

# WORKING WITH "WHAT IS"

"I find no fucking reason to be calm," the kid in the juvenile hall yelled at me. We were sitting in a circle with about twelve other 14-to-16-year-olds in the Brownsville area of Brooklyn, and I had just guided a meditation.

In the meditation I must have encouraged the kids to be calm. I could not recall exactly what I had said, but this kid did not take it very well. "I have so much going on in my life," he continued. "I've got a court hearing tomorrow, there are people who want to kill me, and you skinny-ass-white-mother-fucker come in here telling me to be calm. I find no fucking reason to be calm!"

I sat there, frozen, this skinny-ass-white-motherfucker was speechless. In the meditation teacher training I had completed years earlier at a Buddhist center in northern California, they had never covered how to respond to situations like this. The training was mainly focused on well-behaving attendees asking polite questions about meditation practice. I was not prepared for this work, but I figured the kids would teach me what I needed to know.

The kid stared at me intensely, as if challenging me: "*How are you going to respond to me asshole?*" He did not want to be told how he should feel, and he was letting me know loud and clear. I was getting attacked, and deservedly so. The beauty of working in juvenile halls is that I always knew where the kids stood with me. There were no fake compliments or silent stares. They spoke their minds, and this kid was not holding back.

I actually don't remember what I said to him that day, probably something not very helpful. But his words have stayed with me. Am I telling people how they should feel—you should feel grateful or feel happy or feel calm—or am I directing people to be with their experience?

We are often told how we should think and feel, but much less how to be with our experience. We say, "Don't be angry," or "Don't be upset," or "How could you be frustrated with me after all I did for you?" But our experience is our experience.

**We experience what we experience. Our liking or disliking, our approval or rejection, is secondary.**

Some years ago, the renowned Tibetan Buddhist teacher Mingyur Rinpoche guided a meditation at a Wisdom 2.0 event. He encouraged people to relax different parts of their bodies. Then he said something profound: "If you cannot relax, okay. Allow that you cannot relax. When you allow that you cannot relax, that means you are relaxing."

It is not so much about changing our experience, but changing our relationship to it. If there is tension, how do we meet that tension? If there is sadness, how do we relate to that sadness? No matter our experience, how are we relating to it?

**When we experience tension, it is often not the emotion or feeling that is the issue, it is our resistance to it.**

If we switch, as best we can, from "should" to "what," there is more space and possibility; not what we *should* feel but *what* are we actually feeling? I have learned that whenever I want to tell someone, "You should just relax and calm down," it is generally far more helpful to ask, "Can you tell me more about what you're experiencing at this moment?" By doing so, we build the muscle of awareness.

As Rinpoche suggests, can we allow ourselves to have the experience we're having, to be present with what is? Curiously, when our resistance to "what is" lessens, more space appears. As soon as we stop fighting it, there's often a release. It sounds so simple, yet it's so easy to forget.

# TRIGGERS

"Well, you are an asshole," I snapped at my friend. I was triggered, and after the words left my mouth, I couldn't believe what I had said. The details of our argument do not matter, but needless to say, I was triggered, and instantly felt embarrassed the moment the words came out of my mouth. In that moment, one part of me took over, and I reacted through name calling, as if I was a kid on the playground.

For most of us, moments like this are inevitable. We react and resist, we fumble and lose track of what matters. These are hard moments, and it's also almost impossible to grow unless we are willing to explore and be curious about these moments. If we want to learn, our unconscious patterns often reveal themselves through how we react and get triggered.

Those closest to us, such as family members, partners, and friends, often trigger us the most. A random person could say to us, "You look weird," and we would probably shrug it off. But if our partner or a family member said the same thing, it can trigger a strong emotional response. We respond, "How could you say that? What do you mean? Well, you look pretty weird too."

Initially, it can seem like bad news when we get triggered and lash out, but if we become curious and investigate, it can be great news. Life is giving us an opportunity to learn about ourselves.

**There is something underneath that is now coming into awareness.**

While "holy shit" might be the first reaction to noticing an unconscious pattern, from another perspective, the fact that we can see it means we have grown in capacity. It had possibly existed for years below awareness and now here it is for us to see. Embarrassing, yes, humbling often, yet I am not sure how else we learn.

**Triggers can be opportunities that say, "Learn from me!"**

How do we know when we're triggered? We feel different. Our body has an intensity of sensation and heat that is different from our non-triggered state. Thoughts come more forcefully and quickly, and tension in our jaw and hands often increases. There is usually a strong sense of "other," and the need to protect ourselves. We lose access to inner spaciousness, and our response is intensified. Whatever action we take in these moments is not likely to be helpful, as it is fueled by an unconscious reaction. In many ways, "we" are not taking action, our unconscious is.

At these times, we can inquire: Is my response coming from a place of clarity and ease within me? Or is it coming from tension and discord? Usually, we know the answer. When our response comes from a place of hurt, the emotion takes over, and there is a particular intensity to our reaction. It's "charged" in a way that feels different. In that moment, we've become disconnected from our deeper nature.

The difference in our response lies in awareness. We can be furious at someone, but if we bring awareness to it, we hold that anger thoughtfully and wisely. We feel the emotion with awareness, knowing it needs to move through us. Our job is to be with it, experience it. In this way, we can communicate clearly and directly, without blame or harsh words. We can say, "I'm furious right now," with awareness.

As our capacity for awareness deepens and we become more present, more aspects of ourselves come into the light. Some of what is revealed will be beautiful, like a deeper appreciation for life, and some will be challenging such as long-buried pain or habits we've ignored, from denial to greed, to anger, to hatred—the whole human condition of emotions. It is all in there.

Life wants us to learn and grow. How can that happen if we don't bring to light the areas that were once hidden? So while shame or embarrassment might be our first response to recognizing something uncomfortable about ourselves, perhaps the second is, "Congratulations!"

The key is to meet ourselves with kindness and compassion. When something comes into awareness that was previously hidden, it's great news. It's a sign of deepening. Growth happens when we make the unconscious conscious—the parts that are easy to see and the ones that are more difficult.

I still get triggered by people, and do not always know why. But when I become curious, and inquire, "Wow, isn't it interesting that this person is triggering me," I have space for learning. I may not know the root of the trigger at that moment, but just the willingness to be open to the possibility of learning shifts my experience.

# NON-SEPARATION

A s we deepen our awareness, we begin to notice patterns both in ourselves and in others. As we see patterns in others, it's easy to think, "I am so much more conscious than that person." We start to believe we're further along on the path and that our role is to help others, regardless of their readiness or interest. This mindset, however, often creates a sense of separation, where we see ourselves as distinct or better than others. We then feel compelled to "save" or 'help" others by pointing out everything we perceive as "wrong" or "needing correction" about them.

We might say to a friend, "You know, John, you have this issue with anger, and here's what you should do..." or "Ann, I need to talk to you about this pattern of blaming I've noticed; it really needs to change." We often genuinely believe that they will respond, "Wow, thank you, that is so true!"

But our advice generally does not land quite like that. We may be well-intended, and there may be times to confront people, but in general it is very hard to get people to see what they don't have the capacity to see. If they had the capacity, they likely would have seen it by now.

The fact that someone cannot see a pattern means their system is still developing the capacity to do so. Our thinking that we know what is best for other people or that we need to save other people is understandable, but not helpful.

The goal is not righteousness or one-upmanship, or acting as if we know and they do not; the goal is connection, or non-separation. From a place of separation, we think, "I am *this* and you are *that* and I know better, as I am a more conscious person. Here is the correct and incorrect way to do, well, everything."

We may not actually say that we know better, but we think and believe it. And we may in fact see patterns that are true. But in separation there is a sense that we know more and are better than others. We give people our advice whether they want to hear it or not.

**Rather than asking, "How can I get someone to see something I think they should see?" A more fruitful question might be, "How do I build capacity in myself and support others in building their capacity to see clearly?"**

It may feel momentarily good to believe, "I am better than other people because ..." fill in the blank. We may answer it with "I am smarter, healthier, stronger, kinder, wealthier, more conservative, more liberal, more self-aware" ... you name it. We feel a sense of superiority, and look down on those we see as less evolved. We can never just grab one quality out of humanity and say, "This one quality is what makes people better than others, and I just happen to have it!"

You could be a college professor who identifies as "intelligent" and subtly judges all those without as much knowledge, or an anarchist who judges those who are a part of the capitalist system. The difference between simply acknowledging differences and "separating" is the sense of being better, more evolved, more esteemed. One way or another, we separate.

We can also do this in the other direction. We see from a "less than" lens, believing we are less than others due to various conditions.

From a place of connection or non-separation, on the other hand, we are all on a learning journey. We know people are doing the best they can, and maybe most importantly, we know that we contain every possible quality we see in others. Every quality we admire or detest in others also resides inside us. It's not "John's greed," it's the greed we all share. It is not "Susie's wisdom," it is the wisdom we all share.

**It makes a huge difference when we know that we also have in us all human emotions, feelings, and tendencies—anger, hatred, shame, greed, love, passion, wisdom, and more.**

When we see this, we relate to people differently. We see how we relate to others is how we relate to that part of ourselves. When we see our interconnection, as hard as it can be sometimes, we can have greater empathy, connection, and compassion ... not just for them, but for ourselves too.

# FLEXING

"Well, I've known him for about 20 years. We're really close friends," I responded to a new neighbor I had just met. I was determined to prove how long I had known this particular well-known person, driven by a desire to present myself as important—or at least more important than my neighbor. He took my bait and started doing the same, and back and forth our conversation went.

"Do you know so-and-so? I do," he said.

I responded, "Well, I know him too, but do you know so-and-so? He is a close friend of mine. We are very tight. I was just talking to him yesterday in fact."

And the neighbor would respond in kind. The two of us were in a duel, competing over who had more status, using "who we knew" as our weapon of choice.

After the conversation, my partner asked me, "Why did you guys both feel the need to flex in that conversation?" We weren't flexing muscles; we were flexing who we knew and for how long. Like dogs trying to establish dominance by barking, we were both trying to one-up each other through relation-ships and influence. Embarrassing to admit, and true.

Flexing can take different forms. When I think of men flexing their muscles to decide who is the most worthy and important, it makes me laugh. But somehow, flexing to show how impor-tant I am through the people I know seems natural. But flexing is flexing, no matter the weapon of choice.

**There's no vulnerability in flexing, no real honesty, no sense of togetherness, it is me in competition with you. And where does that get us?**

In Silicon Valley, flexing is common. We flex by name-drop-ping, flaunting wealth, or highlighting our status within a company. The more that power defines or dominates a culture, the more flexing tends to occur. But what lies beneath this behavior? Often, it's a sense of unworthiness, loneliness, or a longing to belong. Flexing frequently masks the tender, vulnerable parts of ourselves.

I flex when I want to feel accepted, worthy, honorable. Instead of recognizing that these qualities are inherent in me, I think, "What is my status in relation to this person? Am I above or below them?" Essentially, I posture. It's no longer two humans on the journey of life having a discussion; it's "How does who I think I am compare with who this other person thinks they are?"

**Status can be derived from one's job title, educa-tion, the size of one's house, or salary—or even how many meditation retreats a person has done. We can flex "spiritually" just as much as it can flex with muscles.**

You can see this at parties or gatherings in the subtle ways we compare ourselves with others, determining who is greater or less than. Yet underneath you'll simply find two humans on a journey, both with struggles and pain, both with thoughts of unworthiness and despair, both with some level of tension in their lives.

The more we let our vulnerability show, we see our commonality more than our differences. We choose connection over rank.

# LENSES OF PERCEPTION

"Have you ever killed someone?" the young kid in juvenile hall asked me with a straight face.

There were about ten of us sitting in a circle on plastic chairs at a juvenile hall in the south Bronx. I was there teaching my weekly class, and the kids were sharing various challenges they were facing. They were all kids of color from the poorest neighborhoods in NYC, dressed in facility jumpsuits, and as usual I was the only white guy in the room, dressed in jeans and a t-shirt.

I looked at the kid, frozen, not quite sure how to answer. I had never even been in a fight, much less killed anyone. I sat there, assessing the moment. Was he actually serious? I was waiting for him to jump in and say, "Man, I'm just joking with you. I know you have never killed someone. Come on, it is obvious. Look at you. You have clean clothes, you wear glasses, you have no visible scars. I know that you have never seen real action. I was just messing with you."

Instead we looked at one another, eyes locked, and I realized he was absolutely serious. I hesitated for a while, then responded, "No, I have never killed anyone." He nodded his head, looked around the room at the others, none of whom seemed to want to chime in.

While there were different types of kids in NYC juvenile halls, most had witnessed, endured, and at times also caused enormous amounts of physical violence and other forms of suffering. They could be in for everything from truancy to murder. In their short lives, many had been in fist fights, gun fights, robberies, recruited by gangs, you name it. From this, they had understandably developed their own worldview.

This kid existed in a world where asking someone if they ever killed someone was a legit question. It was common enough to ask people like me, who were just visiting, as I might ask someone if they had gone to college.

This kid had his own way of defining what was "normal," which was radically different from mine, but I realized that none of us are all that different. We grow up with norms, assumptions, ways we view the world, and then think others do as well. The tendency of course is to think that our conditioning is the right or normal conditioning.

**It is hard to see that our way of seeing is just one way.**

In certain areas, knowing which fork to pick up first at dinner or how to fold your napkin is a really big deal. That helps determine the knowledge and nobility of a person, often not realizing that 99% of the world really couldn't care less. But it matters to them. It is how they base their assessment of someone.

Byron Katie likes to say, "No two people see the same tree." We see the tree through our different lenses based on our conditioning. In the same way, no two people live in the same country or have the same president. We live, essentially, through our thoughts.

When people see the world differently from us, do we judge and criticize, and think, "How could you think such a stupid thing?" Or can we shift, try to understand, and see that our worldview and another person's worldview have equal validity, have equal need to be honored? Can we become curious?

The norm in one neighborhood is to ask about killing and the norm in another neighborhood is to ask what private school a person is attending. They are just different ways of seeing the world based on our conditioning. When we understand our view is simply a reflection of our experience, we can better make space for the many other ways of viewing as well.

We see the world we have been conditioned to see. And the moment we see that, we see a different world.

# PART IV

---

# PATHS

There is no "right" journey or path.

We each get to navigate it for ourselves.

We have to listen to what is inside, to the invisible, to that which does not have a name.

# OUR PATH

Years ago while in my early twenties, I found myself in a large room in Lucknow, India, surrounded by Europeans and Americans dressed in white, their faces glowing with reverence as they gazed at an Indian spiritual teacher seated on a slightly elevated stage. I had come to India seeking to explore a different dimension of reality. Growing up in the United States, a culture largely driven by material gain and capitalist motivations, I felt drawn to something beyond that. At the time, India was a destination for those yearning to explore a spiritual realm, and I was eager to experience this journey for myself.

Here I had hoped to find an authentic spirituality, possibly a profound teacher who could guide me, and teachings to help me expand my understanding of what really matters in life. I thought my understanding of spirituality would grow and deepen from time in Asia. This search led me here on this particular day.

I found a spot at the very back of the room where I could barely sit on the floor because it was so crowded, my knees touching. The teacher spoke for about 30 minutes on the nature of reality, but it all seemed somewhat disconnected to me. I had grown up in a small town on the plains of West Texas, and I had never been in a scene like this. I was also uncomfortable with all these people gazing so devotedly at this man. When the teacher asked if there were any questions, people came up to say how the teacher had transformed their lives, with very few details. One after another adored and praised him saying "You are amazing. You have changed my life." It was deeply touching ... and utterly nauseating to me. I was growing weary of it.

I raised my hand and made my way to the front to sit at his feet, as the other people before me had, to ask my question.

"I have been sitting here watching all these people adore you, and I want to better understand what you are actually teaching?" I asked.

"Where does the 'I' develop in your various sense doors?" he asked me.

It was a reasonable question, and one that really did not have an answer. It was more of an inquiry. But I didn't really get what he was asking. "So what will happen if I see this reality you speak of?" I inquired.

"Once you see this, you will not have this question any more," he followed, as people in the audience giggled.

"You mean, my consciousness will be exactly the same as it is now except this one question will not be there?" I followed, just to be difficult. "That seems odd."

We went back and forth for a while. We were completely missing one another. I had an edge that was in part a reaction to the "spiritual scene" I had entered.

Eventually I went back to take my seat, and the next person came up. "I got it, Papaji! I see it now," the woman said.

"Amazing! How did you get it?" he inquired. "I spent so much time with that other guy and he still could not get it. How did you get it when he couldn't?"

Laughter erupted in the room and people looked at me in the back, smiling and laughing.

For that day, I was "the guy who could not get it." I was not just your average loser, which is never fun, but I was in fact a "spiritual loser," which may be one of the most severe. It is kind of like a "super loser." Whatever "it" was that the teacher was talking about, apparently I was so spiritually deficient I could not see it.

I went to India seeking spiritual teachings, and I thought it might be some epiphany into the nature of reality, a mystical vision, or possibly a feeling of bliss throughout my body.

Instead, I was laughed at.

I may not have received the teaching I wanted in India, but I received the teaching I needed. And it was this: Go to any spiritual place you choose, dress in white or special attire, mingle with those who identify as spiritual seekers, and what do you discover? The same mind. The mind of comparison and judgment doesn't vanish with a change of clothes or location.

Of course, traveling and changing locations can sometimes be useful, but the real work isn't in any particular place. It is with ourselves, no matter our clothes or location. The trip to India was well worth that discovery.

# LIFE AND LIFE SITUATIONS

"The last year has been a year from hell," my friend told me, and he went on to share many of the challenges he had faced.

While this could have been anyone, this particular person is one of the leading entrepreneurs of our age. From the outside, he has possibly one of the world's most ideal life situations: he is enormously rich, wickedly smart, can live anywhere he wants, is respected in his industry, and well-known in the larger society. While that is his outer life situation, he, like all of us, faces challenges that neither money nor fame can shield him from.

I grew up believing that certain life conditions guaranteed happiness. If you had a big house, a nice car, a high-paying job, stylish clothes, and could vacation in Hawaii, you had it made. Happiness was assured! Over time, I realized it's more complicated than that. I met billionaires who had all this and more, yet they were unhappy. Conversely, I met people with very few resources who seemed quite content.

Now, if someone asked you, "Tell me the life conditions that would make you happy, and I will give them to you," many might respond, "If I had a lot of money—say, tens of millions—that would solve all my problems. I could buy whatever I needed. I would certainly be happy then."

Now, let's say your wish is granted. The money arrives in your bank account one day. Suddenly, you're rich! At first, you think, "At last, I've made it. All problems solved. Now I can finally afford all those things I've always wanted: the new house or car, the dream vacation, the best clothes." You can help out friends and family, dine at any restaurant, and hire people to handle tasks you dislike. You have made it!

However, while the money opens certain opportunities, it also brings its share of challenges. Now you need to figure out what to do with your wealth. Do you keep it or share it with your partner or family? You might want a new house for yourself, but your parents or siblings may feel you should buy them a house or a car first. Why keep it all to yourself? And now that you don't need to work for money, what do you do with your life? Do you quit your job? If so, who will you spend your days with? Will you stay at home by yourself with no community?

Sure, you can travel, but that might get boring. What happens if people find out how much money you have? They might try to rob you or ask to borrow money. If you have kids, people may treat them differently, knowing of your wealth. Managing the money becomes a new responsibility. You never worried about finances before, but now you find yourself checking your accounts regularly and watching the stock market anxiously. Before, you had so little to lose; now, you have much at stake. Suddenly, you have problems you never anticipated.

Money can also lead people to develop dependencies. Someone who smoked pot only once a week due to lack of funds now can indulge all day. The money feeds whatever addictive tendencies were already there. As this happens, some people even think, "I wish that money had never come to me. My life was better before."

Of course, not having money brings its own challenges. Many people struggle to pay rent and know that one car accident or hospital visit could erase their entire life savings. They live under immense stress. In some parts of the world, just getting enough food to eat is a daily hardship. Money can certainly help alleviate these struggles.

**All life situations are problematic. They all have challenges.**

No matter our life situation, challenges find us. If we have a body, if we have people in our lives, if we live in a city or town, if we're impacted by the weather and the seasons, there will be challenges. There may be better life situations than others, but there is no life situation without difficulties.

What really matters isn't so much the particulars of our life situation but our connection to life itself. Are we connected to the power of our heart? The inner source of compassion, joy, and resilience?

**When we're connected to life, we feel a sense of aliveness, presence, and vitality, no matter our circumstances.**

The air doesn't care about our life situation, nor do the sky or the trees; all are available to each of us equally. The breath we take, the smile of a child, the joy of laughter, all here, freely provided, regardless of our wealth or status. Of course, it makes sense to create as healthy and comfortable a life situation as we can, but what's even more important is what life itself offers.

No matter our life situation, possibly the most important question is: "Are we connected to LIFE?" Are we alive and present in this moment, the only moment we have?

# OPENING AND CLOSING DOORS

I n the early life of my son, we lived in a trailer in northern New Mexico. It was old, hard to heat, and filled with old thrift store furniture. I wouldn't call it a comfortable place to live. As a result, we spent a lot of time venturing in the nature around us, hiking, seeing friends, and going on day trips in the community. Because the house was unpleasant, we explored the outdoors often.

Years later when my son was a teen and I was making more money, we got a much bigger home with a pool, sauna, and large community room. I thought, "Wow, now we have it all. This is great!" And it was great on many levels, but we also spent more time at home, not venturing into nature or spending time with friends. Home was too comfortable. With that much comfort, why leave?

The new big house opened some doors and closed others. My son was also more likely to invite friends to our trailer than he was to the big house oddly enough because he felt uncomfortable having kids see our big house, as he feared they would treat him differently.

I have a friend who made a lot of money in the tech world and would take his kids all around the world to travel at any school break, and finally the kids revolted, and said, "Can't we just stay home and be with our friends here? Why do we always need to be going somewhere?"

Yes, they had the money to travel the world, but how are the kids supposed to form deep friendships when they are continually spending their free time on a jet to some foreign land? So is it a blessing that somebody can spend their summers and breaks traveling the world with their kids, or is it a curse because they don't develop relationships with those in their area and have a sense of place?

**Money opens some doors and closes other ones.**

I have seen again and again that the families with the biggest homes often have siblings with the weakest bonds. Why? Because kids are often in their own room on their own devices, not connected to one another. Kids in smaller homes who have to share rooms have much more contact, learn to get along with each other, and tend to share more about their day and get closer as a result. They are somewhat forced to get off their devices and connect.

Of course, very small homes have their challenges too. I grew up in a small 3-bedroom house with 5 kids, and it was very chaotic at times. It was hard. I dreamed of someday having a big house and my own room. I prefer the challenges of having money to not having money, but everything opens doors and closes doors.

To me the most important question is, "What is beyond the doors? What can we rest in that is independent of the doors?"

# DIGESTION

"What have you been up to for the last few hours?" my partner asked me. I paused, realizing I didn't have a good answer. I knew I had been looking at my phone, scrolling through social media, conducting some Google searches, but the specifics eluded me. "Just looking at my phone, I guess," I finally replied.

Why couldn't I recall anything meaningful from the past several hours? Despite consuming vast amounts of content, from texts to videos, I lacked full attention and presence. In those few hours, I likely encountered hundreds of posts, yet none left a lasting impression.

I'm guessing I'm not alone in this. Today, we have countless ways to take in information: books, podcasts, videos, social media, and more. But if I asked you, after three hours of scrolling through social media or watching various videos, what you had learned, could you tell me much? Particularly when consuming content through platforms where posts appear and disappear rapidly, it's challenging to retain or reflect on what we've seen.

**Our challenge today isn't a lack of answers. We have solutions for nearly everything. The real challenge often is in taking the time to truly digest those answers.**

It is one thing to hear an answer, it is another to listen deeply and actually digest an answer and to convert it into wisdom. How many amazing, wise, funny, insightful posts have you seen on social media that you cannot recall now? The words were consumed by you, but not actually digested. Your mind was simply racing to the next thing. In some ways, it is not just the words, it is how we digest those words.

The information we consume has the potential to be transformative, yet instead of giving it time to settle, exploring its depth, savoring its subtleties, letting it resonate in our hearts and memories, we swipe, scroll, and move on to the next message, then the next. The issue isn't always the quality of the information, or the "food" we're taking in; the issue lies in our ability to absorb and digest it fully.

**When our minds are constantly active, jumping from one post to the next, information lands like seeds on concrete.**

There's no space to absorb what we're taking in. It's like trying to enjoy a nutritious meal while sprinting down the street. Even if the food is the healthiest and most delicious ever, we're neither savoring it nor properly digesting it. In our world today, we're often sprinting mentally, and find it hard to slow down.

While certain types of content are better than others, it's not the words you read here or anywhere else that truly matter. It's you, and your capacity to be curious. What the best words can do is point toward things you may already know.

So when you turn a page or scroll down a screen, take time to digest. There's no need to hurry. This moment is just as accessible and important as any future moment. When we take a breath and notice the world inside and around us, connect to our bodies, and appreciate the moment, we're able to see and respond to life more effectively.

We learn to consume less, but digest and absorb more.

Notice when you have been consuming content online for some time, how do you feel? Does it help to consume even more content or do you need to stop, connect with yourself, and take time to digest what you have already consumed?

# FROM LOVE OR FOR LOVE?

There are different ways of giving and engaging with others. Sometimes, we aren't fully aware of our intentions. We give, but with various expectations and agendas. We might offer to help a friend with a report, give work advice, or drive him or her to the airport, all while telling ourselves, "I just want to help. I'm doing this from the goodness of my heart."

But later, if they don't return the favor, we think, "I did this for you, and you didn't even...." It turns out our giving wasn't as free from expectation as we believed. Underneath, we may have held an unspoken contract: "I am doing this for you, and in return, I expect you to...." We had an unexpressed agenda.

Of course, most friendships involve natural give and take, and if one person is only taking, the friendship won't last. However, when we find ourselves frustrated by someone's actions or lack of response—we feel like we helped them and they never returned the favor—we can ask ourselves, "Am I doing this *from* love or *for* love?" If it's for love, or some expected return, we will be disappointed when we do not receive the response we hoped for.

If it is, in fact, transactional, it seems only appropriate to be upfront with someone, and say to the person, "I'm doing this for you with the expectation that you will do the same for me. Do you agree?" The friend can then decide whether to receive our help or not.

If what we are offering is genuinely from love, however, then what basis do we have for being angry? We might think, "I did this out of love, with no expectation for anything in return, it was simply out of the goodness of my heart, and in return you didn't even thank me for it." But was it really from love? Was there truly no expectation? If we get angry, we had expectations.

**Genuine giving *from* love feels much better than giving *for* love.**

From love finds satisfaction in the act itself. The giving is its own reward. This is what the world so desperately needs, yet this is also the kind of giving that can be so challenging. It does not help to give as part of a hidden campaign to get more likes, followers, appreciation, or approval. We don't need to show how generous we are. It's about the act, not the image or what we get in return.

Of course, it's nice to be thanked, and giving and receiving appreciation is beautiful. But it's also important to understand our motivations, whether we are acting for love or from love. The most powerful giving is when we are nourished by the act itself, when there is no future outcome required for us to feel at peace; the act is the return.

If there's no natural give-and-take that emerges in a relationship, we can choose to spend less time with someone. We do this not out of righteousness or frustration, but because of a misalignment in priorities. There's no need to tell others,

"John is such a jerk. I did so much for him, and when I asked for something, he wouldn't do it. Now I do not talk to him." No story is needed. We simply no longer feel compelled to engage in the same way.

**Our job is, as best we can, to act from love, and let go of all that is out of our control.**

What hidden agenda motivates us as we engage with others? Is it transactional? Thank you's, kind acts, compliments, and picking up the check at lunch ... all can have an agenda. They can all be strategies, all means to an end. And the end is often appreciation or attention. It is not really giving, it is more of a business transaction.

So, yes, it is great to be with people who appreciate both giving and receiving, but if we truly take responsibility for our choices, we notice what hidden agendas may be driving us, and enjoy the act for itself. When we connect to that deeper source, the gift is the reward.

# KNOW YOUR MIND

I f you could deeply know one thing, what would you choose?

You might answer that you want to know about physics, technology, or AI. Of course, it would be cool to be an expert on numerous topics. You could impress your friends, be a popular guest at dinner parties, and likely make a lot of money.

However, while knowing all these topics might be helpful, possibly the most important of all is knowing your own mind, understanding what creates suffering and what brings ease. Imagine having all kinds of knowledge, but not knowing the inner dimension: not knowing how to sit quietly by yourself, how to get a good night's sleep, how to listen and connect to another person.

A meditation teacher I knew once traveled to Burma (now Myanmar) in the 1970s to study with the great masters of that country. After practicing for some time, he was granted an audience with the foremost master in his tradition. He was told the master had very little time and could answer only one or two questions.

When he met the master, he asked the most essential question he could think of: "What is the essence of your teaching?" The master responded immediately, "To know your mind."

Seeking deeper understanding, he asked, "But why? Why do I need to know my mind?"

The master replied, "For the benefit of all sentient beings, know your mind."

I have always loved this response because it holds both a personal and a collective truth. We know our mind for ourselves—but not just for ourselves.

**The mind is the forerunner of all actions. The greatest acts of love originate from the mind, as do the greatest acts of hatred.**

Every word we've spoken (and that has been spoken throughout history) that has hurt another person originated in someone's mind. And every act of care and kindness in history also began in someone's mind. Every time we've felt a sense of belonging and love, or a feeling of disconnection and isolation, it was an experience rooted in our mind. While external elements can influence us, ultimately, our experiences begin and end within us, with our mind.

When we do not know who we are, do not understand our mind or heart, it becomes challenging to navigate life effectively. We keep thinking other people are the cause of our problems, that other people "make us feel a certain way." We never see the role our own mind plays. Thus we keep creating the same scenarios again and again.

Of all the things we can know, understanding our mind, heart, and inner life might be paramount. This is the secret of people in almost every position, be it business, politics, sports, and even the best ultimate fighting champions. One of the greatest fighters, Khabib Nurmagomedov, said, "People watching my record and say that this guy is tough. This is not about tough; this is about mind." Be it in Myanmar or Dagestan where Khabib is from, it is the same lesson.

So when asked, what is the most important thing to know? I do not know anything more essential than our own mind.

And how much time in a day do we set aside to explore this?

# UNDER AND OVER AWARENESS

W hen I was younger, whenever I felt anger, I often buried the feelings. Rather than acknowledge or express what I felt, I kept it bottled up. If someone had asked me if I was angry, I often replied, "Nope, all is good." I simply wasn't ready to face or admit what was actually going on inside. It was simply not an acceptable emotion.

However, I would often silently resent the people who had hurt me. I would be subtly critical, or make side comments. Essentially, I would close my heart to them, all while pretending everything was fine. Outwardly, my stance said, "I'm fine," but inwardly my state of mind was, "You pissed me off, and I'm going to make you pay for it in ways you may not even notice."

The pattern of silent resentment when I was hurt we could say was "beneath my awareness." I couldn't see it at the time. This is likely true of all of us. There are aspects of ourselves we're aware of, and other patterns we don't see. They are "under awareness," much like the depths of an ocean hidden beneath the surface.

When we notice our anger and understand its pattern, we could say that it enters, "the frame of awareness." We can feel the pattern in the body, and notice the thoughts in the mind. We get to know the experience of anger. We are familiar with it. Anger is there, and so is awareness.

However, if we suppress the anger or only recognize the impact after we lash out at someone, that anger remains "outside the frame of awareness." When patterns are beneath awareness, it is like having an invisible co-pilot in our car who occasionally steers us in unexpected directions. We don't know how we changed lanes or ended up in Dallas instead of New York; we simply find ourselves there.

You might be asking, "So how do I see something I cannot see?" It's a great question. While we can't force awareness, we can create conditions that allow us to see more clearly.

One way we miss our patterns is by surrounding ourselves with people who share the same tendencies. If we have an addiction to alcohol and spend most evenings with people who drink heavily, it's hard to see our addiction. Our drinking looks and feels normal. Same with families. If everyone in our family yells during conflicts, and we adopt this pattern, yelling seems natural. Awareness often arises when someone mirrors back to us something different.

Practices like meditation are meant to expand our frame of awareness. Silence or spending time in nature can have a similar effect. We create space "to see what needs to be seen and feel what needs to be felt." However, if we constantly watch TV, scroll through social media, and hang out with people just like us, it's challenging to see beyond our conditioned ways. How can insight or clarity emerge when our attention is swiping from one post to another every few seconds? The mental traffic is so intense nothing else can enter.

**A simple practice throughout the day is to explore "open awareness."**

While standing in line or walking down the street, gently ask yourself, "What might it be like to expand my awareness in this moment?" Notice the world around you, the colors, the shapes, the people.

Practice bringing awareness to different moments in your life, especially when "nothing is happening." Look up, feel your body, notice what is true in that moment, take in your inner and external world. You are not trying to make anything special happen. You are simply awake to the world around you.

Awareness builds capacity, and capacity supports clear seeing.

# WHO IS RESPONSIBLE?

W hile one tendency is to suppress, another is to hold on and blame. This happens to me at various times. When I notice myself in a shitty mood, for example, often my first response is to find someone to blame. I think my mood is because of something someone else did: my friend stood me up, my partner was rude to me, or my son was not appreciative of my help ... or any number of other reasons. I believe the thought, "It is their fault that I feel the way I do!"

Now, you might direct your focus differently, and blame your manager for speaking harshly to you or your child for throwing a tantrum. We all have our targets, our usual suspects. No matter how we direct our blame, the general vibe is, "If these people had not done what they did or these things had not happened, I'd be fine. But since they did, I'm forced to feel this way."

**We believe that the world dictates our state of mind, that external events determine our inner experience.**

Of course, they may somewhat. We are naturally sad when we experience loss and naturally angry when we experience something unjust. These are important. Sometimes it is almost impossible to not be impacted. If, for example, someone is giving you electric shocks every few minutes or putting your hand in a scolding hot fire, that shit is going to hurt, bad. Life sucks at times. And there is not a lot we can do. A loss or hardship can put us in deep mourning for some time.

AND we also have some impact on our state. We have agency. Two people could have an upsetting event, both lose their job for example, and it impacts one person for a day, and someone else 20 years later is still talking about that event! "Man, I should not have been laid off. Those assholes." He has carried the stress of that event for 20 years!

The same event happened, but they had very different ways of relating to it. So yes, the situation matters, but so does how we hold and "are with" the experience.

Two people could have 9 positive events, and one difficult event in a day, and one person has gratitude for the nine positive events, and the other laments about the one negative event. I would not say one perspective is "good" and the other "bad," but one comes with much less stress and pain than the other.

Now, nothing needs to be suppressed or denied. That just creates more dis-ease, but we can learn to be with all that arises in a way that allows it to move through. And we can notice the temptation to create stories around our experience, "This should not have happened to me," or "If only this person had acted differently" or "I have to suffer because of them."

These may be true at one level, but does believing these thoughts help to ease our state of mind or does it create more suffering for us? We are not blaming ourselves or anyone else, we are simply inquiring, "Who is responsible for our current state of mind?"

Then we can see what answer we get.

At times it can seem as if we are learning something new.

Other times it feels like we are remembering something that we always knew.

But then we forget again.

And then we remember.

What did we forget?

Exactly!

# THE NEXT MOMENT

I had just finished a big event that went exceptionally well. People enjoyed it, and I received a great deal of praise. But the next day, I felt a sense of loneliness and emptiness. "What is happening?" I asked myself. I had just experienced an incredible day. It did not make sense that I would feel this way afterwards.

Then it struck me, "I know what I'm experiencing." I thought, "I'm experiencing the next moment."

The next moment arrives after any pleasant event, be it the most delicious meal, the greatest sex, the best vacation, or the most engaging conversation. There is the event, and there is the feeling we have after the event.

Everyone has a next moment. It doesn't matter how wealthy, successful, or important we are; the next moment always comes. Every meal ends, every vacation ends, every love-making ends, every applause ends. External stimuli have a pattern: they arise and they pass. When external excitement fades, we're often confronted with our inner world and this can be uncomfortable.

When we're fixated on having a good time, the next moment can feel like an enemy. We rush from one thrilling experience to the next, trying to outpace the inevitable lull that follows. After one party, we're planning the next; after one great dinner, we're scheduling another; after one trip, we're already dreaming of the next. We believe that constant stimulation is our only path, fearing any quiet moments when it's just us, without any external stimuli. As a result, we run from ourselves, and from life.

**If we want to live more freely, however, the next moment becomes our friend and ally. It can show us how much we are or are not connected to ourselves.**

Once a positive external stimulus is over, what do we notice? How do we respond? Do we immediately plan our next trip or meal or movie? Or do we pause and listen to what's present when there's nothing to entertain us? And do we feel grateful for what has just happened? If we're on a path of learning, awakening, and deepening, the next moment is our teacher. If we are content and comfortable with who we are, we welcome all of it, the moments of excitement and the moments of no excitement. It all has beauty.

After the great vacation or trip, there's the return home. Did our experience change us? Perhaps, but here we are back in our familiar surroundings, with just ourselves again. Of course, we may have new clothes, fun experiences, and some great memories, but fundamentally, nothing much has changed; we're still ourselves. And the harder we try to run from the next moment, the more challenging it is once it catches us, which it always will.

**When we derive our sense of worth from life's circumstances, the next moment becomes both feared and sought after.**

No matter who we are, the next moment arrives. It whispers, "Enjoy all the journeys of your life, but know that in the end, it's just you with yourself. You can don new clothes, attend great parties, indulge in various experiences, but ultimately, you cannot escape yourself. So you may want to get to know this amazing being that you are. This is home base."

There's no way to avoid the next moment, but we can choose to learn from what it has to teach us. No matter what happens, the next moment finds us. Do we welcome it or push it away?

# THE INHERENT

Like many people, I enjoy acquiring things. I'd much rather live in a spacious, comfortable home than a cramped, noisy, or dirty one. A new car or new clothes bring me a certain sense of pleasure. I value quality items and take joy in the finer things in life. Yet, at the same time, I recognize that all of these have their limits.

There are so many objects we can acquire—land, homes, reputations, power, followers, and more. In fact, much of life can be consumed by this single pursuit: acquire more! It's understandable, given that those who accumulate the most are often celebrated in society. "Look how much that person has!" we exclaim, marveling at their planes, homes, yachts, and prestige. We elevate them to the pinnacle of success, placing them on society's highest pedestal.

Yet, if you read about many of these individuals, you discover that often, the more they have, the more miserable they are. Family members sue one another, magazines publish unkind stories about them, and they have to deal with people constantly asking them for money.

They also know that their status is fragile. They can be up one day, down the next, perpetually on the treadmill of more success and fame, coupled with the fear of failure, of who they would be if they lost this status. This often makes them extremely insecure, as they know themselves only through what they possess.

When acquisition isn't the source of our identity, when we know who we are with a little or a lot of items, we see that it's beautiful to own land or a home, and it is also beautiful not to own anything. We see that whether we have one house or four, there's something far more important: the inherent. This involves discovering something that already exists within each of us regardless of how much we acquire.

Of course, it's not really about the houses or possessions; it's about the identity. We may think, "I do not own any home and only have a few possessions. I am so much better than all these other materialistic people."

No, you are just turning the game around, still missing the inherent. You are just believing a different story. From the place of the inherent, none of this matters.

Possessions often trap us in the belief that they hold the key to happiness, blinding us to their limitations. The money we passed down to our children might be used wisely or it might foster a life of entitlement and aimlessness, as they live off inherited wealth without ever having the opportunity to carve their own path.

What we thought was medicine can become poison. "If I can just give them this, they'll be happy," we think. We fail to see that while possessions have their place and money can be used for good, there's something far more vital: connection to the inner dimension. There is no better inheritance. When we see that the inherent exists equally in the poorest and the wealthiest among us, the rest is play. Since it can never be lost or added to; nothing we do gives us more or less of it.

Imagine coming to the end of life and, as you say goodbye to everything you've acquired, you realize that you know almost nothing about the inherent. You ignored the inner dimension. You achieved so much—your name is on hospitals or university buildings, you have a bank account with impressive numbers—yet something is missing.

**The dance is to engage with the world of "things" but to also know that the real game is inside, it is in discovering the inherent.**

We can see this in children that are given everything, the best technology, the best education, the best vacations, yet miss what they need most: the presence and care of their parents. The parents cannot offer what they didn't learn themselves: the connection to the inner dimension.

When we access it, we can offer it.

# THE MAKING OF IDENTITY

I t is easy when growing up to feel more or less than others. We get molded into a role or identity.

We have a B average in High School, and someone else has an A average, and we experience a sense of less than. Then we meet someone else who has a C average, and we feel a sense of more than.

We graduate from a respected state school, and someone else graduates from a lesser known college, and we feel a sense of greater than. We then meet someone who graduated from Harvard or Stanford and we feel a sense of less than.

We get a job as a manager, and someone else gets a job as a secretary and we feel more than. And someone else our age gets a job as an executive and we feel less than.

Over time, our "sense of self" starts to harden. We get to know the *level of somebody* that is appropriate for us. We look at LinkedIn profiles and other social profiles, and the hardening continues further. We see the other accomplishments of our peers, and compare ourselves accordingly. "This is who I am" we begin to believe. "I am better than all these people and less than all these people. This is my status in the world."

Of course, there is no one agreed upon criteria for status. Everyone makes up their own: for some people, size of house matters; for others size of social media following matters; and for others power and influence matters. For some physical strength and beauty is a major factor, and for others it is mental intelligence.

We choose our metric, which often means we choose our way to feel more than or less than.

We feel less than or more than depending on who we are with, and how we measure up to them. We often stay in our safe social circles where we can feel equal or better than others, avoiding situations where we feel less than, while often subtly both judging and craving other people's situations. Essentially, we believe that we *are* our status.

To a certain extent, this is inescapable. We're social animals. But the dance is to play in multiple dimensions, similar to Jesus' encouragement to be in the world but not of the world. In the world, we may have some status or not have status, but we are not only of this world. There is another dimension, and in that dimension, there is no separation, nothing that can define us as better or worse. We are everything. It is timeless and formless.

**We are the deeper intelligence, as is every other being no matter their station in life.**

And when that becomes our resting place, we walk through the world differently. It is not that we do not see status, it is that we know who we really are regardless of it. Instead of getting hardened, we learn to soften. We can lose objects, houses, cars, titles, our reputation, but who we truly are can never be lost.

We see both the relative (the world of form) but also the absolute (formless, timeless Being). We know who we are, so we can play in the world more effectively.

Real wholeness can never be added to or taken away from.

# DRAW A LONGER LINE

I magine what our life would be like if we spent less time trying to control things that we cannot control? Complaining is one important area to explore.

I love to complain as much as the next person. Sometimes it feels good to let off steam, express my judgments, and not really give a shit. The danger lies when we get stuck in a cycle of complaining, and it becomes a constant. When this happens, no matter what's happening, there's always something wrong. We think, "Someone is doing something they shouldn't," or "Someone is not doing something they should." The weather isn't to our liking, nor are the stock market trends or our level of popularity. Sometimes, we can't even pinpoint the source, but there's a lingering feeling of dissatisfaction.

We think:

> **"Something is wrong with this moment, and I am unhappy about it. If I stay unhappy about it long enough, maybe it will magically change!"**

A Zen teacher once went up to a chalkboard, took out a piece of chalk, and drew a line on the board. She then asked her students, "How do you make this line shorter?"

Some students suggested erasing part of the line from the top, others from the bottom, and still others proposed erasing the middle of the line. After each response, the teacher shook her head. Finally, after more guessing, she took out her chalk again and drew a longer line next to the first one. "This is how you make the first line shorter," she explained. The original line now appeared shorter simply because the longer one was next to it.

There are situations (the first line) that need to be changed and addressed, if we can, and there are times when trying to change the first line does not help. We need to draw a longer line. Sure, our boss or our friend can be a pain in the ass, and we can complain about them all day and night, but at some point we can ask ourselves, "Instead of changing them, what if I focus on changing myself? What might it mean for me to draw a longer line?"

**We generally cannot control what happens in the external world, what comes our way, but we do have some choice in how we respond via our inner world.**

The root of our dissatisfaction of course isn't the weather, the market, or our popularity; it's our relationship to the present moment. Complaining focuses our attention externally. It has a particular energy behind it; we dislike something, but we also enjoy our disliking. We enjoy being the person who was wronged. The victim role feels comfortable to us. It allows us to stay stuck, and justifies us viewing other people as the problem.

But as we notice this pattern, we see more options. We realize that we can draw a longer line. The next time you find yourself complaining, maybe inquire, "What if I draw a longer line? Instead of trying to change the situation, might I change myself?" Or we can leave the situation, if possible.

But the energy of complaining drains our energy, and there are so many ways we could use and harness that energy to be of service to ourselves and to humanity. As the saying goes, "It is better to light a candle than complain about the darkness."

# SEVEN MINUTES

"I'm trying to make it to seven minutes," a friend said recently at dinner, as we discussed morning routines.

"Seven minutes for what?" I asked.

"Seven minutes before I look at my phone in the morning. It's hard, but I'm trying," she said.

Of course, no one is forcing her to check her phone the moment she wakes up. Her boss is not telling her that if she wakes up at 6:30am, she must check emails by 6:37am. That is not the case for her or most of us. Yet, like many of us, she simply feels compelled to do so. Like her, our phone summons us, and we often heed the silent call.

Why do we feel this urge? I do not know. Perhaps it's an ancient instinct to seek out information that helps us feel safe, reassured that no danger is imminent. We check the news, social media, or emails to confirm there's no emergency. But it is a dangerous way to live if we need to look at our phones every five minutes asking, "Am I safe now? Is everyone else safe?" It becomes our addiction.

**With our phones today, there can be almost no time to be, to non-do.**

We can rationalize it, thinking we need to be next to our phone at all times, because, well, what if an emergency were to happen? What if our friend needed us, or our child or a coworker? The thinking is, "At any moment something could happen, so I need to be available at all moments."

But humans survived for hundreds of thousands of years without phones and alerts, and likely will do just fine in the years ahead. Just like my friend, the truth is that it does something for us.

**Essentially, we feel that we must always be doing and avoid non-doing.**

However, non-doing is essential for our nervous system. It gives us space to settle, to just be. As writer Pico Iyer said, "We are living at the pace of machines, instead of the pace of life, and this is a very hard pace in which to live." Our body works differently than technology, at least at present. It needs ease and spaciousness. We cannot live at the pace of machines.

A musician I heard about required his audience to arrive 30 minutes early before a concert, explaining that they needed that time to let their minds settle so they could truly take in the music. Rushing in five minutes before the concert would be a waste; they wouldn't be able to fully enjoy it.

Giving ourselves space from technology, be it the first 30 minutes or last 30 minutes of a day, is a great act of sanity.

When we pause, we can actually listen, both to what is inside us and what is around us. The quieter we become, the more we can hear and see.

# THE FEEL

"**D**id you check your email while you were in the bathroom?" my partner asked recently during our date night. I usually don't check my phone when we're together (well, sometimes I do), but she was right this night, I had. I justified it by thinking, "Well, I'm not really with her while I'm in the bathroom, and it doesn't take any extra time away from her. Maybe something important came in. So it doesn't matter, right?"

It turns out, it matters a lot. She could feel the difference. It's easy to notice when someone checks their phone while you're together, even if it's just for a moment; the connection shifts. We are no longer fully present. Of course, there are times we genuinely need to check our phones, such as an urgent call or text. But most of the time, it's not truly necessary. We're just running on habit, not fully aware of the impact.

It's fascinating to observe how, in a group, when one person picks up his or her phone, others often follow. Test it out yourself. It only takes one person to trigger a chain reaction. Before we know it, often everyone at the table is staring down at their phones.

What's even more interesting is that people can sense when you've mentally entered the digital world, even when you're out of sight. My partner could feel it when I checked my email in the bathroom. Why? Because I brought that mental processing back with me to the dinner table. My mind was still transitioning from digital back to analog, and this can take time. My mind was still processing the emails or texts I had read and she could feel it.

> **Checking our phone doesn't just take up the moments we spend looking at it. It also requires time to transition back into the analog world.**

The cost of these distractions is not just time, it's a shift in our attention and presence. Checking social media or email, even briefly, introduces new stimuli, whether it's excitement from an interesting message or frustration from an unwelcome email. We need to process these emotions, and it takes time to transition back to the analog world.

Thich Nhat Hanh liked to say, "the greatest gift we can give another person is our presence." We each have this gift, and get to choose how we offer it.

# QUALITY OR IMAGE: YOU CHOOSE

I was walking in a state park recently, and saw a peculiar but common sight: scores of people taking pictures of themselves in front of a beautiful ancient redwood tree. However, from what I could tell, no one was actually looking at or connecting with the tree. It could just have easily been a picture of a tree. The actual tree mattered little.

It's as if everyone thought, "Oh, a photo in front of this tree will get a lot of likes. This will impress people." And off they went, focusing their phones and tapping away to get the right photo, but missing one of the true wonders of this world, an old-growth redwood tree. They may have looked good in the photo they took and shared, but it appeared as though they missed the goodness available at that moment.

**We often face a choice: be fully present in the moment or to try to capture the moment with a photo.**

In a world dominated by screens, it's easy to prioritize image over quality.

Journalist Anderson Cooper shared a story with me during an event in NYC. "No one ever really wants to talk to me," he said. "Even when I introduce myself, saying, 'Hi, I'm Anderson. What's your name?' most people just want a selfie." He explained that people want the social media clout of showing they met him, but they have no real interest in engaging with him.

"You see these people on Instagram, at the beach in Mykonos, and you wonder if they're having such a great time, why are they stopping to take a picture?" Anderson asked. "It's all about exclusion, showing everyone else who isn't there that you're having a better time than they are."

I love taking photos too, but the real question is, Is it intentional or habitual? Does it bring us closer to or further away from the present moment? Is it an expression of joy or distraction?"

I have seen influencers posting in ambulances on their way to the hospital, seemingly about to die. And what is on their mind at this time? Taking and posting pictures of their journey in the ambulance and possible demise. In such a dire and important time, with so many loved ones that could be reached out to, this is their focus.

They seem to be thinking, "How might this tragic situation help my brand? Yes, I might die, but maybe in the process I could at last be more relevant? My followers might increase." As far as I know, though, there are no social media accounts allowed after death.

Of course, there are moments to take pictures, but if we miss our life, it is a big price to pay.

# BULLSHITTING BEING

W e can bullshit a lot of things.

We can bullshit that we are smart by reciting what other people have written as our own, bullshit that we are rich by taking pictures of ourselves in front of other people's cars, bullshit that we look a certain way by using an image editor or filter. There is so much that we can bullshit.

Yet thankfully there are some things that we cannot bullshit about.

There was an indigenous tribe in South America in which, when someone committed a serious crime, they were forced to walk across a field in front of an execution squad. The only chance of survival was for the condemned person to walk in a way that expressed integrity, so that no one could pull the trigger. The tribe believed that if the guilty person had acted from a deeper truth, if they had committed the crime to be helpful to the society, they should not be punished, and this would be revealed in how they walked across the field. The person who was motivated by truth would walk differently than the person motivated by hatred or anger. The firing squad would sense

SOREN GORDHAMER

this, and thus no one would pull the trigger and the person would go free.

I've always appreciated this story because it highlights the power of authenticity. When someone is in an aligned, open state, you can feel it. No amount of posturing or external presentation can replicate that. We all emanate our inner world. When we're truly present and acting from a place of truth and integrity, others feel it. And when we're disconnected, lost in our thoughts or stories, they feel that too.

**No matter how hard we try, there is no way to pretend or bullshit presence. We are either connected to ourselves and the moment, or we are caught in some strategy.**

We can meet someone and think, "Wow, this person has really got this presence stuff down." The person wears trendy clothes, has read all the spiritual books, has traveled to Asia and the Amazon jungle, often pauses in conversations to appear thoughtful, wears beads from India, and even has a picture of himself with the Dalai Lama or some great spiritual teacher, which also serves as his social media profile. It is all there and looks perfect … yet something is missing.

And that something is presence or Being.

We tend to think, "If I can get the external form right, purchase the right clothes, my internal will change too." And it may, a little, but it is far more direct to do the inner work. Fortunately, there is no pill, no quick fix, and no easy path to Being. We cannot buy it or inherit it. We cannot bullshit Being.

With nowhere to go, nothing to do, and no one special to be … what is left?

# THE VULNERABLE ASK

J amal was frustrating me. He came to the voluntary class I led at the Juvenile Hall in Brooklyn every week, but did not engage very much. In the class, I guided the kids in meditation, simple yoga, and held a circle where we shared experiences. Most of the kids put in some effort, but Jamal seemed disinterested. He looked around during meditation and barely participated in yoga.

However, after every class, Jamal always came up to me, thanked me, and gave me a hug. It was not a long, mushy California hug, but a quick New York hug. Still, it was a hug.

Over time, however, I grew frustrated with him. I thought, "Why do you come if you do not participate?" I wanted him to try harder, and could not understand why he didn't. It was a voluntary class after all. No one was forcing him to be there. I just did not get him.

I thought, "Why are you wasting my time and your time by being here, if you are not even going to participate in the class? What's your problem?" I would say good-bye to him after class, but I was always a little distant.

Then it hit me one day: Jamal wasn't coming to the class for meditation, he wasn't coming for the yoga, he wasn't coming for the group conversation. He was coming for a hug after the class. That is what he needed. He didn't have a problem, I had the problem. Why should I care if he came to class for a hug? In fact, Jamal wanted that hug so bad he was willing to sit through a class he wasn't interested in just to get it! That was real dedication.

My guess is there is a Jamal in all of us. It is hard to ask for what we need. It is hard to say, "I need a hug." Of course, we may not get it, but there is a time to ask, and let it be known what we are seeking or needing.

How often do we ask for advice, when we are really just needing the support of another person? The non-verbal request might be, "Yo, lets hang out," and the vulnerable request might be, "I am pretty lonely right now. Any chance you want to spend time together?"

Both requests have their place, of course. I am not saying vulnerability is always the best choice, but it certainly is at times. In my experience, men tend to be awful at this. We often see vulnerability as a form of weakness, so we don't ask for help. We have a huge loneliness epidemic in our culture, especially with young men.

And if we are afraid to acknowledge our loneliness, and ask for help, we stay stuck. We spend hours and hours a day playing video games, watching TV, scrolling social media mainly to distract ourselves from the fact that we are lonely. It turns out, we need one another.

If we can begin to see it and make space for it, we can make vulnerable asks. Brené Brown writes, "Vulnerability is not winning or losing; it's having the courage to show up and be seen when we have no control over the outcome."

Of course, vulnerability can be terrifying, but with roughly one third of people reporting they are often lonely, we are not alone in asking for help. Sometimes all it takes is one person reaching out with a vulnerable ask, and in that vulnerability a connection can be made.

What if we make space now for a moment of nothing much happening.

We will be bored for a minute, together.

Nothing stimulating or thought-provoking here.

Just this.

Just life unfolding right now.

# CAPACITY AND RANGE

"I don't care what you do, just do your best," I remember telling my son when he was younger. He had a strong interest in basketball, and I encouraged him to be the best he could be. Secretly, I wanted him to succeed, to win, to be a star player, and I can only imagine that this came through my voice when I spoke.

Is telling a kid to play his or her best a positive message? Maybe. From another perspective, why not just play sports for fun? Why not just go and goof around and have a good time? Who cares who wins? Instead of, "Do your best" why not say "Go have fun"? Or "Go do whatever the hell you want."

We tend to think of strengths and weaknesses. It is a strength to work hard, and a weakness to not work hard or goof around. I see them as the same: we have strengths, and these same strengths in certain situations are weaknesses. The real question is more about capacity and range.

Imagine a child who is told to work hard and she gets all kinds of stars and praise for her achievements, in sports, academics, arts, you name it. She learns that this is what matters in life, and gets accustomed to the approval she receives when she succeeds. She becomes a winner. Yet if this is the only muscle she develops, if we fast forward 30 years, my guess is that this same person has a lot of issues around stress, high blood pressure, and anger.

If this is her focus, she likely works 60-hour weeks, does not take time for herself or her emotional life, and has a hard time sleeping and making friends. She is so busy achieving, so busy working hard. She is likely not going for walks, chilling out with friends, enjoying vacations, and making sure she has plenty of time for rest and fun.

So is working hard a strength or weakness?

It depends on the situation. Rather than focus on our "strengths," I think it is better to develop capacity and range, to know when it's time to surrender and when it's time to fight, when it's time to work and when it's time to relax, when it's time to talk and when it's time to listen.

**When we have capacity and range, the moment calls us to action, not the stories that we were given about how we should be or what should matter to us. We listen and engage organically based on what is needed.**

Hard work may be a strength at one time, and a great weakness at another time. From the perspective of capacity and range, we can call on what is needed in each moment, allowing it to guide us.

Do we have the capacity to be patient when patience is needed? And act forcefully when the moment calls for it? To be quiet and listen when the moment calls? And to speak up and be heard when needed?

The answer to all these questions is our ability to attune to a given moment. Everything changes. The so-called strength that got us where we are today may be a weakness as we move forward. The answer is in our capacity. If we tune into this moment right now, what does it need? And can we build our capacity to meet it?

That is it.

# PART V

## THE CREATIVE

The invitation is to both attune to what is here now … and to also tend to and follow what, if anything, calls us.

When we are less encumbered by our stories, we have more space to create.

# WHAT WANTS TO HAPPEN

It is easy to keep busy, and spend our days "doing" just about everything—posting, clicking, reading the news, scrolling through social media, you name it. Even in the little free time we may have in a day, we can fill it with constant activity. There is nothing wrong with keeping busy, but to be creative, we need to drop into ourselves more deeply. Rather than live in a constant state of "doing" and activity, we can make our actions or "doing" focused and intentional.

In the process of deciding what to do or what actions to take, one question we can ask is, "What do *I* want to do?" This is always an important question. I mean, why shouldn't we focus on what we want? We can contemplate what truly matters to us and how to follow that more fully in our lives. Super important.

However, over time, I realized that this question has its limits. It harnesses a particular intelligence. Another question we can ask is, "What *wants* to happen?" One is me asking a question to myself, and the other engages another power, a universal intelligence, a mystery.

Both have their place.

We can make a decision from our intellect, thinking, "This is the right decision. It makes sense to me and likely will to others too." Or the decision could simply come from intuition; it feels like something that wants to happen. When we start adding layers, thinking we're better or worse, attaching stories that boost pride or stir up shame, the decision takes on a different quality. It's no longer a process of intuitive listening. When intuition is involved, the rational mind has a seat at the table, but it is not in charge. Something else guides.

At times in our life, it can be helpful to inquire, "What wants to happen?" and then listen. Often our body knows more than our mind does. And if we take time to sit and listen, to be with the unfolding, as if we are in a type of dialogue with the universe, we can know at a deeper level. The space says, "Nothing needs to happen, but if something wants to happen, we are open." We listen, as if guidance is coming our way.

**From the space of not needing, so much is possible.**

Another force comes through that is both us and much deeper than us. But that force needs to know that our primary focus is not the "thing," not the project. It is instead being fully present with what is. As we take time to listen—we sit in meditation or spend time in nature or go for a walk—often we can hear this voice more deeply.

Since we are fine with "just being," we can "do" with the same ease and spaciousness. We are not hurried and stressed, we are not trying to fill an empty pit inside us. We are in what we may call spacious or conscious doing. We are not trying to be anyone special. The completion of an action does not make us better. We already know who we are. We do not need any life event to determine that.

As we listen and inquire as to what wants to happen, we can feel a pull toward an action and we can be equally fine with nothing special happening. Like a surfer in the ocean, we do not try to surf when there is no wave. The moment with no waves is wonderful too. And when the wave comes, we let it take us. The wave has the power; our work is to get out of the way and let it take us.

# THE MIRAGE OF SOCIAL MEDIA

"I don't actually use social media much at all. I go on, post, and then get immediately off," a friend told me recently. She has millions of followers across platforms and has used social media to build a thriving business. But she limits her time there and uses it only because that's where people's attention is.

I often smile when people get excited because an influencer responded to their comment. "I can't believe they actually replied to me," the person says, feeling a sense of connection with the influencer. But, like my friend, most influencers aren't sitting with their phones, personally responding to random comments. They have teams of assistants for that and increasingly AI bots. The influencer isn't scrolling through the comments; they're on a yacht in Greece or in a studio creating their next project.

**The most creative people are, well, creating.**

They limit their time on social media as much as possible. Sure, it may be fun for a moment, but no great work of art or significant innovation has ever emerged from scrolling through 60-second videos all day. True creators know this, and they structure their lives accordingly. They want to use their limited heartbeats on the planet to make an impact, and endlessly scrolling through posts does not help them.

Most of us could benefit from the same strategy.

Use social media for your purposes, but don't let it use you. It can be a source of entertainment, but we get to decide how much time we spend there and whether that time is truly enriching. The most creative people I know use social media simply to share information, and they spend as little time as possible there.

It is not about watching others, it is about creating what we are called to create. If you want to take the most important lesson from the smartest and most effective social media influencers: limit your time on social media, and focus more on what matters to you.

# THE CALL

E ach of us is born into a unique life journey. None of us are here by accident. There is a purpose, a calling, waiting to unfold within us.

Often, our early conditioning steers us onto a familiar path. And this was important for a time. We follow what is expected and do what we are told. However, eventually we often feel called to step into the unknown. This call asks us to listen to an inner voice that is not based on our past conditioning and family story. It arises from a deeper, more authentic source.

There can be different motivations or energies behind an effort. We could start something as a way to avoid something else, or out of a sense of lack, fear, anger, or a wide range of emotions. Anything can pull us in a certain direction. However, when we really listen, usually there is a deeper pull, something "meant to be." It does not have to be a big deal or special project, but it comes from this intelligence inside us.

**We notice where our energy is drawn. It is not so much something we *should* do, but something we *must* do.**

When we step out of our story and see more clearly our life conditioning, we can get closer to what is true and real inside us. There is no image or identity that we are trying to meet. We are not looking to prove ourselves, and we are not bound by the rules of the culture.

When we step beyond the "shoulds," we touch an intelligence beyond societal norms. It is a subtle, inner knowing, an openness that guides us, if we take the time to listen. This intelligence often lives in the subtle: it speaks through instincts, body sensations, and the whispers of our heart.

The form, essentially, does not matter. This could involve spending 40 years in corporate America, working at a restaurant, or caring for animals. There's no single vocation that defines, "This is what it means to be aligned with your calling." It's whatever calls to you, whatever within you seeks expression in a particular way. The money it earns or the fame it brings might matter somewhat, but something else matters more: the place it is coming from.

When we tune in, we can hear another dimension. The inner is guiding the outer. And this is trustworthy.

# IT'S ALL ENERGY, MAN!

"I've never said this publicly before, but it's all energy," my friend said. He's not an energy healer or meditation teacher or Tarot reader—he's a successful business leader who sold his last company for tens of billions of dollars, and has been on the forefront of business innovation throughout his career. He has led tech companies of various sizes to great success. "Most entrepreneurs aren't ready to hear that," he continued, "so I have to hint at it. But it's all energy."

What does he mean? To me, he means there's an unseen field we can tap into that others can feel. Words are important, but it's not what we say that matters most; it's how we say it, how we carry ourselves, the excitement we bring, the depth at which we live what we say. This all determines energy.

For example, if you apply for a job you think you can do, you can say, "I can do this job" with a sense of confidence, and people will feel that confidence. You look the interviewer right in the eye, talk clearly and calmly. You are not trying to impress or get something from the interviewer. The person can feel your capacity. You know there will be challenges, but you have the strong sense that you will be able to meet and overcome those challenges. You cannot say for sure, but your attitude is, "I can do this, and if there is a challenge, I will figure it out."

Or, you can say those same words, "I can do this job" with fear and hesitancy. The energy behind your words is entirely different. You do not make eye contact, you say the words quietly or under your breath. There is a sense of hesitation and uncertainty. You may be having memories of past times you have failed, and those impact you in the moment. You say the same words, "I can do this job," but the person interviewing you hears, "I am not sure if I can do this job."

This is not about acting confidently; it's about embodying confidence, connecting to that part of us that is confident, and trusting our capacities. It is not pretending; it is more about accessing. And if we cannot access confidence, we need to ask, "What part of me is not sure I can do this? What part of me believes I am unworthy of this?" Clearly, something is getting in the way. This is just as true in an interaction with a colleague, or neighbor, or friend. Our energy matters.

**No one will believe in your value if you don't believe it yourself.**

You are not trying to prove something. You do not care what others are doing, or who the other job applicants are. You do not really even care if you get the job. That is not your business. You are connecting to something deep inside, and you are sharing that. You are letting the rest unfold as it needs to.

**If you carry the sense that you are worthy, that a company would be lucky to have you, people will feel it. It is not arrogance, it is confidence from the inside.**

Life is energy and energy is life. When we are connected to this energy, our actions are superpowered. We are not pretending, we are simply accessing that truth in us, and doing what we need to release whatever gets in the way.

Are there certain elements that help harness that energy? Of course. It may help to research the company you are interviewing for, have the experience needed for the job, and other related preparation. If you're applying to be a physician without a medical degree, confidence will be hard to come by. There may be essential skills to develop and work to complete before you're truly prepared for that role.

But no matter how much work you have completed, you can always connect energetically to the moment. You can say, "I have never done this before," or "I would have a lot to learn, and I am excited to start," or "I know this is a big jump for me, and I like challenges." The person interviewing will feel your authenticity. You may not get the job, but you are learning one of the most important skills: connection to your experience, and the ability to express that authentically.

The rest we really cannot control. Whether we are liked or not is really none of our business. Rick Rubin put it this way: "Success comes when you say, 'I like this enough for other people to see it.' Other people liking it is out of your control." This is true for art and true for life.

When we are aligned with the energy, in a very real way the energy does the work.

# ENERGY BEFORE IDENTITY

"I would love to do this," said my friend, "but I'm a corporate executive, not an artist." Another way to explore a calling is through the lens of energy. In this case, my friend's energy was pushing her to create art, but her identity said, "That's not me!" There is a conflict between the story in her head and where her life pulls her.

I am guessing that this is true of all of us at times. The mind says, "This is me, and this is not me." But the energy could care less.

However, could my friend be more than she thinks she is? Could she trust a deeper source as she navigates her life? The mind tends to separate, while the energy just knows how it wants to flow. Sometimes we have to choose between the energy we feel and the identity we hold.

One of the guiding principles I like to follow is "Energy before identity." If our energy is pulled in a particular direction, even if it doesn't match our identity, it's best to trust the energy.

**Energy doesn't care what we've done in the past, what others expect of us, or what degrees we hold or do not hold. It's communicating on a different frequency.**

If our energy wants us to start a company, but our mind says, "I'm not an entrepreneur," it's time to expand our view of who we are. The energy is showing us something emergent that needs attention. If it doesn't fit our identity, it's time to expand our identity, no matter what it is. The energy cannot be argued with. It is what it is.

I enjoy writing these days, including this book, but if I think, "I am a writer; I must do this," or "I am not a writer; I cannot write," the process of writing changes. If my energy isn't there and I write only out of obligation, people will feel that. Similarly, if I believe I am inadequate or unworthy of spending time writing, this will inevitably hinder my progress. All we can really know is whether we have energy or not for something.

When we get disconnected from our energy, we lose connection to life and creativity. We go through the motions of an old identity, even though we have no energy for it. It becomes something we *should* do and "makes sense outwardly," but it may make no sense internally. Worse, it attempts to get others to validate something hollow in us, to reinforce an identity that ultimately does not serve us.

We may fulfill a role, and get accolades for doing so, but in many ways, the role is empty. It might have everything … except our life energy! Of course, at times we need to fulfill commitments and do things we don't enjoy, and sometimes only for the money. However, at some point, when we are ready, we realize that the energy is more trustworthy than any identity we created for ourselves.

How do you follow energy toward an area of interest? I think it helps to first acknowledge this as the truth of our experience, then see what is workable in our current life situation. Sometimes this means committing to the activity that calls us for five minutes a day, then letting it slowly build; or it could involve telling a friend what truly calls us and having one person who can support us as this new path develops.

**When choosing between what you think will have the greatest impact and where your energy is most drawn to, follow the energy, suggests author and legendary producer Rick Rubin.**

We tend to the energy as best we can in the way we can. Now, you may respond, "How is five minutes a day going to do anything?" Five minutes a day creates a relationship with that part. We are giving it attention, nurturing it as best we can. We are saying, "I see you. I hear you." We are building a connection. Then "it" begins to talk to us, be with us. It is alive.

Even if we cannot fully follow something at this time, we can still partner with it, listen to it, and see how it might more fully emerge. The important shift is moving from "something that will happen later" to "something that is happening now."

Five minutes a day of creating art, or writing, or making music, or building that company, or whatever makes our heart sing ... all of the sudden something starts to come to life.

# LACK + FULLNESS

"What are we going to eat and where are we going to sleep tonight?" I asked my friends one day as we walked across Pakistan. We were deep in the north, traversing the Karakoram Highway toward China. Calling it a "highway" at that time was certainly a stretch—the road was barely paved.

Around us, I could see nothing but barren wilderness: sheer cliffs, towering peaks, and the occasional trickle of a stream. To my question, one of my friends wisely suggested, "Instead of thinking there's no food or shelter around, what if we think there is food and housing around and we're just not seeing it?"

As odd as this suggestion was, it shifted my perspective. And this became something of our mantra on the walk: "What if what we need is already here, and we're just not seeing it?"

At first, the idea seemed ridiculous. After all, I couldn't spot any sign of food or shelter in sight. But as we embraced this mindset, I found that it transformed how I approached the unknown. It shifted my focus from seeing scarcity to recognizing possibility. Over the three months we spent walking across Pakistan, this shift proved increasingly helpful. For the roughly 90 nights we slept outdoors, shelter appeared more often than not.

Sometimes, that shelter was as simple as spreading our backpacks under a tree at night. On some occasions, "food" meant little more than drinking water from a stream, but more often than not, food and shelter revealed themselves. We could more readily see food and shelter options from a place of fullness instead of lack.

There are different ways to walk through the world. We can live from a place of lack, from feeling less than and inadequate, as if we are missing something; or from fullness, as if we are complete and life is presenting what we need. Of course, telling someone who is starving that there is actually food around is cruel and unhelpful, but there is a place for mindset.

I used to have an issue with the word "manifest." People would say, "I manifested this watch I am wearing," or "I manifested this job I have." First, there are so many conditions that determine whether something happens or not. Maybe your thoughts had some impact, but I'm not sure that means you can take full credit for anything.

Second, if you can actually manifest something, how about manifesting no more school shootings? If you can manifest a watch, surely you can manifest no more hunger in the world, right? Why not use your thoughts to manifest these things? Why just focus on yourself when there are so many needs in the world?

I have shifted somewhat since then. I have come to understand that our mind impacts our experience. If you are in a job that you hate, it is powerful to begin to envision a job that you love. In the first line of the Dhammapada, the Buddha is quoted as saying, "The mind is the forerunner of all things."

On the one hand, if we walk around believing thoughts such as, "I will never be happy," or "I will never get a good job," the chance that either will happen is extremely unlikely. We first have to believe something is possible. Without an openness to possibility, that possibility has little chance of taking form. We have little chance of noticing or responding to possibilities that arise.

However, if we want to be a successful artist or actor, how do we think that moment will feel if it arrives? And can we feel that now? Rather than thinking, "This thing I want is not here now, but someday it will be," we can act as if what we want is already here. That way, we live from fullness instead of lack. If we want an effort to succeed and we think, "Once this happens, I will feel successful," we will be disappointed. We are coming from lack, waiting for life to fill us.

**What if, instead, we carried ourselves as a success now? What if we did not look to the future or the outside world to determine our sense of worth? What if we lived as if what we desired was here, then let life do what it needs to do?**

From lack, we tell ourselves: "There is never enough. So many people have it better than me. If only I had made this other choice in the past everything would be fine. If only I had invested in Bitcoin or Nvidia 5 years ago!" But this thinking stops us in our tracks. From the perspective of fullness, there is a world of possibilities, of opportunities, and it is more about tuning into a particular frequency in ourselves.

How do we do this?

We start to shift from "it's not here" to "what if it is here?" We open to possibilities. Then, as we get closer to that frequency, life naturally responds. But life is not here to try to fulfill us, to satisfy us, to make us feel worthy or important. We have already solved that. We know this in ourselves, and in carrying ourselves as enough, as worthy, as whole, when external circumstances later appear that reinforce that sense, it feels natural. The world mirrors what we already feel inside.

# SHOW UP

"I'm going to do it today," Navarre, my five-year-old son, announced at an indoor pool with a five-story-high slide. He was a new swimmer, and the slide was intimidating.

"Dad, I'm headed up," he said as he started up the stairs. Minutes later, he was back by my side, still dry. "I couldn't do it," he admitted. "It's too scary."

This went on for twenty minutes—Navarre climbing up, then coming back down without sliding. Each time, he said, "I wasn't ready. Let me try again."

Finally, I saw a flash of blond hair as Navarre came swishing down the slide, beaming. He splashed into the water, swam over, and said, "That was so much fun! I'm going to do it again!"

I stopped him and asked what helped him overcome his fear. What was it that at last helped him have a breakthrough?

"Well, Dad," he explained, "I was standing at the top, and it was my turn, and I was just about to turn back again, but I slipped. And before I knew it, I was going down the slide."

"You slipped?" I asked, flabbergasted. I had hoped for a more profound answer, maybe about my encouragement. But no.

"So you never decided to go down?"

"No, I slipped. I'm going again now!"

Then I realized that what is true of water slides, is often true of life. At times, all we have to do is keep showing up.

Often, we face situations where we think, "I don't know what to do. I do not have an answer." We can avoid showing up or making a movement due to our lack of knowledge or an answer. However, the answer (or answers) at times only arise once we are in that moment. We might not know beforehand, but we can know in the moment, if we have the courage to show up.

Ahead of time, we think, "I am not sure what I will do. I need to know." But curiously, if we show up for that moment, and allow ourselves to "not know" and we don't make a problem of not knowing, "knowing" can happen naturally. What will we say to our manager if she is angry with us? What will we tell our child if he loses the game? If we show up, generally the answer will too.

**We may not know, but it is possible that the moment, once it arises, does know.**

It is hard to know ahead of time, but if we have a willingness to show up, to listen and become curious, often there is a shift. There will be many situations where we won't know how things will go. The problem isn't the not-knowing; it's our discomfort with not-knowing. Not-knowing or "beginner's mind" is often the place of creativity. When we can allow it, something magical can happen.

We can take that uncomfortable meeting, give that talk that scares the shit out of us, ask that person out who may reject us, take that job we think might be past our ability.

If we show up, often the answer we need will show up as well. We keep showing up until we slip.

# LOOK BETWEEN THE LINES

F rom spending decades interviewing and getting to know some of today's leading entrepreneurs, one quality almost all of them share is the ability to see between the lines.

They see the normal world, but they also see more ... while everyone else is only looking at the words on a page or images on a screen, they also see the space around the words and images. They have the ability to shift into a different perspective.

The world says, "Look here," and "prioritize this!" and "this is what matters," and "to go there, you must travel this way," but with a slight shift, we can see more possibilities. The creative has more space to play. We do not always shift the situation, but we shift how we view it.

Rules are here to play with, not always to follow. For example, most editors tell me that there are not enough words here to make this a chapter, yet here this is, as a chapter. Does it work? It does not matter. It is what it is.

How might you look at your work from a different lens? If you are accustomed to one perspective, can you take the other perspective? If most people are going right, consider going left. If people see a problem, what might it be to see the same situation as an opportunity?

Just the willingness to ask the question can create a shift.

# TRENDS

F or a while, working at a large company like Google or Nvidia can be popular … *Then it's not. And more people want to be independent entrepreneurs or influencers.*

At other times, fitting into the system, knowing how to "play the game" is popular … *Then it's not. People are more interested in going against the system.*

Certain clothes are popular for a while. There is a look of coolness and success. *Then they are not. Fashions change.*

Playing a sport, be it tennis or golf or a certain video game, may be popular for a while …

This is true of everything. If we watch this pattern long enough, we see everything has a pattern, a flow. What we view as "in" is only true for that moment. Nothing lasts.

Entrepreneurs try to see the patterns and get ahead of them. They want to build the app before apps become popular, create the AI coach before AI coaches get popular. They want to see and join the future, so it does not bypass them. Any business effort has to be attentive to trends.

While some people try to predict the future, a handful of entrepreneurs create the next chapter of society, be it AI models, robots, rockets, or electric cars. They are not guessing or waiting for trends to change; they move before most people even know anything is happening. They have the vision, determination, and tenacity to build the future we will live in.

**In the end, while trends change, our vision and values need not.**

If our effort stands for something, it should not be dictated by the winds of what is popular. We are aware of where the world is going, and adapt as needed, but our inner world still guides. We may adjust our strategy, but not our values.

There will come a moment at the end of our life when we might ask ourselves, "How did I spend the limited heartbeats here? Did I honor what truly mattered to me? Did I live in alignment with my deepest values?" At that moment, the answer, "I chose what was popular or trendy over what was authentic and real for me," may feel deeply unsatisfying.

**While some people chase trends, we can also let trends chase us.**

The dance lies in being aware of trends, and even creating them at times ... while still allowing our inner life to lead. What others are doing or finding interesting or is popular at a given moment doesn't ultimately matter. If we are truly committed and clear about what matters to us, in my experience, we do not need to chase trends ... trends eventually chase and find us.

There is the energy of laziness and distraction … and there is the energy of hurry and anxiety.

There is also the energy of the creative.

With nothing to prove, no inner hole to fill … allowing what wants to come through us to come through.

We do not so much create it, as we allow something to be created.

# TRY SOFTER

Often when we think of effort, we think of exerting force, and certainly that is a part of it. But if you have played sports or competed, you have probably learned and seen that there are different ways we can "try."

One way to try is to think that effort alone will solve our problems. We try with force and tension. We are "here" and we just need to get "there," and the faster we get there, the better. "There" may be losing weight, building more muscles, or growing a popular social media channel. Our goal is to get "there" as soon as we can, and we think the best way to do so is through constant effort and force.

This may work somewhat, but it is tremendously taxing on the body, and it does not always turn out well in the end. There is another way to move through life with intention and focus, not force.

An old story tells of a martial arts student who asked a master how long it would take to become a master himself.

"Ten years at least," the master replied.

The student didn't like that and asked how long if he worked twice as hard.

"Then it would take twenty years!" the master answered.

"Why does trying harder make it take longer?" the student asked.

"With one eye on the destination, there's only one eye left to find the way," the master replied.

**I remind myself of this often. What kind of effort am I putting in? Is it spacious, trusting effort, or am I trying to force something into existence?**

In Silicon Valley, the tendency is often to try too hard. The pressure to succeed can actually impede success, adding tension to our body and mind. How can we make hard decisions or access intuition under such stress? When we exert too much effort, we miss the feelings and opportunities of the moment, miss our ability to access our intuition.

Sometimes as we find our way, what we most need is to pause, breathe, appreciate the moment—to try softer. This kind of trying has a lightness and ease to it. We exert energy, without thinking effort alone is the answer. We are less "trying to get somewhere" and more "letting life unfold." We know when to move forward and when to stop. Our effort is steady, focused, and soft at the same time.

# SPIRITUALITY AND MONEY

"**I** am on the spiritual path, money is not important to me," I remember telling a friend when I was younger. It was more of an identity at the time. I wanted to make those who sought or made money appear "less spiritual" while I was "more spiritual" for my stance. Fortunately, my relationship with money has changed and adapted over time.

There are several truths we need to hold: One, for the vast majority of us we need money to survive, and there is absolutely nothing wrong with making lots of money and having wealth. Second, when we are on our deathbed, the difference between earning $100 thousand or $100 million over a lifetime is unlikely to hold any real significance. It's hard to imagine anyone taking their last breath and proclaiming, "At least I own four houses," or "At least I made it to senior vice president!"

Research consistently shows that while a certain amount of money can improve well-being, having a sense of purpose and cultivating deep, meaningful friendships are far more important. Having more money doesn't make someone a better person, and having less doesn't make someone worse. Some people excel at making money, while others shine in different ways—like teaching children or serving those in need. The fact that money managers earn so much more than teachers is, to me, profoundly disheartening.

**At the same time, money is also energy, and how do we relate to this important aspect of life?**

I have had the great pleasure of working with some of the most destitute and poverty stricken kids in NYC juvenile halls and some of the wealthiest and most powerful people in the world in Silicon Valley. I have learned a lot from each. The real issue is our *relationship* to money, not only how much money we have. Now, if I was in charge, I would create a much more fair tax system, without the loopholes that exist today, where billionaires often pay a lower percentage of taxes than their secretaries do. I think it is pretty clear that the game is rigged for the rich.

At the same time, this is what we have, and our job is to both work for change, and play the hand we are dealt. For years I did almost every job possible: I washed dishes, worked at a cafe, was a cashier, roofed houses, lifeguarded, delivered magazines, and worked various odd jobs. Every job taught me something important. And there is absolutely nothing wrong with doing a job mainly for the money.

I tried to live simply so I had more free time to work on my own projects. When I first started my company, Wisdom 2.0, I lived off about $1,000 a month—$500 rent for my trailer in the woods, and $500 for everything else. I did various odd jobs, then in my free time, I worked on developing the business. Now, if someone else was also developing a similar conference, and lived off $4,000 a month, I would have 4 times the runway of him or her, just by living simply.

Simplicity at times allows time and space for an idea or project to develop. At other times, growth calls us to expand and invest large amounts of money in ourselves or in a project we want to develop. A friend years ago inherited about $20,000 and she asked various people for investment advice. Does she put in stocks, real estate, or something else? A friend of hers said, "Why not invest the money in yourself?" And she went on to create a very successful company that she loves. But for that, she had to put all the money she had in that endeavor.

We need to know when to conserve, and when to go all out, and no one but ourselves knows the answer.

What we are looking to create is a quality life, where money has a seat at the table—not the best seat, but an important seat. We balance the need for resources with our need for purpose and right livelihood.

# THE MONEY DANCE

I like to see money as a dance. Do we resist money? Grab and hold onto it? Ignore it? Spend it instantly because we feel we are unworthy of having it? How do we dance with money?

There are no set rules. I have seen people who lie and cheat and make a lot of money. I have also met people who live with integrity and do not have many resources. Money is not a determination of our worth or impact on the world. In fact, many of the greatest spiritual teachers died with almost no material possessions. On one level, it says nothing about what matters.

AND most of us need it, and there are shifts we can make to live in greater harmony with it.

For example, at the time of this writing Warren Buffet is 94 years old and worth about $174 billion. Yet would you trade positions with him? Would you rather have all his money, but likely much less time to live, or be your current age (assuming you are under 94) with the money you have? Most people choose time over money.

However, there are some patterns we can explore.

A teacher of mine, Lynne Twist, works with people across the economic spectrum, and her exploration is always the same: What is our relationship to money? She once worked with a woman in massive debt. Alongside managing money better, Twist suggested the woman explore how to be generous, even while she was in debt. A dollar a month to a cause, a smile to a stranger, offering time to help a friend, whatever she could.

The woman in debt soon started to see herself differently. She was no longer "someone who needed help," she was also someone who could give help. As a result, she began to carry herself differently, and moved from thinking, "I do not have anything" to "I have a lot to offer."

Over time, her financial situation also shifted. The woman's new orientation allowed her to engage from a place of abundance, and the universe responded. The universe often reflects back our experience, and money is no different.

**Can we be generous when we do not feel generous, access kindness when we feel wronged, love when we feel unloveable? Sometimes the remedy is in going deeper and embodying what matters to us, regardless of what life is giving us.**

If we are generous only when we have money or kind only when we are happy or loving only when we feel open, we are letting external conditions guide our actions. The real invitation, as hard as it is, is to live our priorities regardless of the external, and money is one place to learn. Can we be generous no matter our financial situation?

Are you a good channel for money? If the universe were dispersing resources, and looking for good people to help move money, would you be trustworthy? Are you concerned for everyone, the whole, or just yourself? If not, what changes would need to take place to carry yourself differently?

As we make these inquiries, it often becomes clear how we either resist or constrict with money. If we can let it come, and let it go, and if we can be of service to something deeper, money more often flows to us and through us.

Sometimes the money is flowing in …

Stay the course.

Sometimes the money is not flowing in …

Get another job if you need or find a way to make money, but still stay the course. Stay true to yourself.

# MAKE IT EASY

At Google's HQ, they conducted an interesting experiment: they placed healthy snacks at eye level, while making people kneel down or reach up for unhealthy options. The unhealthy food was there, just harder to get. The results? People chose healthier snacks without ever being told to do so. Google just made it easier to choose that option.

Google could have conducted awareness campaigns on the dangers of unhealthy food, tried to educate people on the benefits of healthy food, or many other efforts, all of which might have helped some. But simply making the choice easier, it turns out, was a direct way of supporting change. This way, no one needs to be convinced of anything, and all the usual choices are still available.

There's a lesson here for us. We are creatures of habit. If our phone is right next to us when we wake up, it encourages one type of habit. If we wake up with a yoga mat right next to us, that invites another habit. Both lean us in one direction or the other.

Of course, no one forces us to check our phone first thing in the morning, but when it's right there, it silently calls, "I might have important updates, want to see them?" It can be hard to resist. If that is not what we want to do first thing in the morning, it doesn't make sense to have our phone right next to us when we wake up. The invitation is too strong, and there is no reason for it. If you say, "I want to spend the first hour of my morning phone-free, but I keep my phone right next to my bed," you are simply making it more difficult for yourself.

Physical spaces make invitations. Walk into anyone's home, and you can see what matters to them. A 60-inch TV screen in a room clearly invites attention. It says, "I'm here, and I'm very important!" It expresses itself as a priority of the person or family that lives there. It's similar if you see large book-shelves filled with books. This says, "Books are important in this house. We value books."

If we look around our environment, what does it say? What are we "making it easy to do" in our physical environment? And we can experiment. What happens when we sleep with our phone in another room and have a yoga mat or book next to our bed instead? How do we then start our day?

In a world full of constant distractions and screens vying for our attention, our physical spaces invite us into certain ways of living. Children in homes full of screens—laptops, phones, tablets, TVs—live more of their lives on screens. The physical environment continually beckons them, "Look at me. Pick me. Spend time with me."

Take a look at your physical space and ask: "What are the items around me inviting me to do?" Does this align with what matters to you? Does this setup express your values? If not, what changes could you make? We are constantly receiving non-verbal invitations, and it makes sense to see if they align with our values.

If the space around you in your life was extending an invitation, what would that be? What are you being drawn into? If you want to support contemplation, for example, what might it be like to have a screen-free room, or have an empty chair or meditation cushion in rooms whose sole purpose is to give you space to do nothing special. Let's make our environments as supportive as possible, so we can do what's hard, but make living aligned with our values easier.

What does our physical space invite in us?

# THE RIGHT KIND OF HARD

There are different kinds of "hard." There's the hard that comes from ongoing conflict, lack of sleep, or working a job we despise. This type of "hard" needs to be addressed, if possible. This hard impacts our body, our well-being, and can prematurely age us. Our biological age might be 30, but too much of this kind of hard wears us down, so we look and feel 50. We are caught in a loop of negative mind states, conflicts, and very little purpose. Life is hard.

Yet there's another kind of hard that stretches us, activities such as meditating for 15 minutes when you've never tried it before, lifting weights, jogging, apologizing to someone for a mistake you made, or asking someone out on a date. These are hard too. They are often uncomfortable. We would much rather sit on our couch scrolling through social media or eating chocolate as we binge watch another Netflix series than go for a jog or do something that challenges us. This kind of hard wakes up our body and mind, it enlivens us through the difficulty.

In the past, success for many people was synonymous with a life of leisure: having enough money to pay others to take care of everything for you—gardening, cooking, walking the dog—so you can relax, watch TV, and surf the internet with no responsibilities. This was seen as "making it." But it turns out that stagnation is one of the worst things for health and well-being. If you don't use your physical muscles, they weaken. If you don't use your mind, it weakens too. I've always appreciated the phrase, "Do only what is easy, and life gets hard, but do what is hard, and life gets easy."

While constant stress is harmful, the right kind of stress is actually beneficial. Our bodies and minds need to be challenged. It's a way of telling them, "You are needed! Stay vibrant. I still need you. Pay attention."

Doing what is hard means going for that run or walk, lifting weights, meditating every day, or having that difficult conversation you've been avoiding. Exercise might be unpleasant in the moment, but it makes life easier in the long run. It's the hard work that eventually becomes easy.

Focusing on what's easy is, well, easy. We all know how to do that: eating sweets, staring at our phones for hours, avoiding difficult conversations that could deepen relationships, shopping to fill a void, or smoking to forget our fears; those are easy. Doing them is living in a sort of trance, avoiding anything that might wake you up, might bring you out of your daze. Without awareness, we simply go from one daze to another.

**In the end, the easy path turns out to be quite hard.**

Look at adults who've lived easy lives, who had all their needs and desires met, who never faced real challenges—they're often not happy. Growing up with servants and trust funds, never learning how to do things for themselves, doesn't create an easy life. It's not easy to live without resilience, without the ability to do hard things when necessary, or without taking responsibility for your own life.

People who avoid challenges often end up facing harder lives. Without developing resilience, they struggle when difficulties inevitably arise.

Many of the most rewarding experiences of my life have been hard. Not hard in the sense that they caused pain or harm, but in the way they expanded my sense of what's possible. When we finished our World Walk in Hiroshima, Japan, much of the satisfaction came from the perseverance it required. Had we just flown to Japan, it would have been a different experience. Walking, with all its challenges, made the journey powerful.

Building a business from scratch, while enormously challenging at times, brought me more confidence and friendships than I ever imagined. When you work on something you're passionate about, people often feel compelled to help. If your effort comes from a place of heart and care, support tends to follow.

**The question isn't, "Will this be hard?" Most worthwhile efforts will be. The real question is, "Is this the right kind of hard? Will I learn from this? Does this feel like my calling?"**

Then the difficulty is fuel.

# GREATNESS

I t's easy to judge ourselves by our current situation: "Am I succeeding? Am I getting more followers? Am I making more money? Am I getting more respect from others?" In many ways, these are quite normal and understandable questions. However, from the perspective of true greatness, these are secondary.

The greatest artists, authors, and business leaders often work in obscurity for years. Sometimes, they aren't acknowledged until after they pass away. They persevere, not because they see results or get recognition, or make money, but because they are driven by a deeper purpose. They are fueled by a calling.

**Greatness comes from attuning to something bigger than just our current results.**

Some podcasts I follow have been around for twenty years. In the first ten years, they had very few listeners. Over time, their audience grew. If you're not inspired by what you do, it is an excruciating process. But if it's your calling, then the journey itself becomes the reward. You may need to live simply, not go on vacation or eat dinner out for many years, but it doesn't bother you, because you are fueled and guided by something deeper.

I was having dinner recently with someone who co-founded and led a multi-trillion dollar company and had a net worth of tens of billions. Yet talking to him, the success and money mattered little. He loved what he did, and woke up every day committed to his team and creating the next generation of tech products. He had spent decades building his company, experiencing both incredible failure and embarrassment, and also great success. But something mattered more: dedication to an effort that he felt passionate about.

Now, I can disagree with people who spend their lives building a particular company or fighting for a cause I do not believe in. I may see it as unimportant or unhelpful, but it is hard to argue with their passion and commitment. These people are exceptional at what they do; they are committed to greatness, and challenge themselves to go beyond what anyone thinks is possible.

Greatness emerges from challenges, learning, and perseverance. The struggle itself helps build the depth and power of our work. Recognition is secondary. Greatness is not about how the world responds to our efforts; it's about how deeply committed we are to our craft, to our life.

We do what we do as fully as we can, and the world does what the world does.

# WHAT IS INSIDE

"Come on Johnny," the parent yelled at his kid at bat during the baseball game. The kid, no doubt, was doing his best to keep his eye on the ball and get a hit, but the tension was palpable. A lot had depended on the play.

The coach paced back and forth, clearly agitated by how the game was unfolding.

Well-meaning parents yelled words of support, "You got this, kid," and "You can get this hit." Still, the tension mounted.

Watching this, it was hard to imagine, with such pressure, how any kid could play his or her best.

Of course, there is nothing inherent about competition that creates tension. There are simply kids on a field playing by certain rules. There is nothing about playing a sport that needs to be stressful.

The tension, of course, was because of something else, not always said, but certainly felt: a desire to prove oneself, to be a winner instead of a loser. It was not the game, it was the thoughts and expectations surrounding the game.

All this was in the background as this child stood at the mound that day, getting ready to try to hit a ball, a hit that might have determined how he, his family, and his teammates viewed themselves.

It struck me while watching the game: this is all of us. We are all up to bat in the sport of life. And how much tension do we create due to pressure we put on ourselves?

Michael Gervais, a sports psychologist, works with competitive athletes. His job is to help some of the world's top athletes mentally prepare for their sport. For years, he also served as the sports psychologist for the Seattle Seahawks.

I asked him one day, "What do you tell a player before he or she heads out to the field or court?" Do you say, 'Play your best' or 'Good luck'? What guidance do you give?"

My son was playing basketball at the time, and I was always a little perplexed about what to say to him before a game.

Michael replied, "I generally say the same thing, no matter the sport or the player."

"And what is that?" I asked Michael.

"I tell them, 'you have everything you need inside you'."

**Winning and losing is not the point, he explained, a player discovering his or her own innate power is a much better focus.**

Then something very curious happens: as a player focuses in this way, often his or her performance improves, as the tension decreases.

It turns out that winning more often happens when something else matters more.

If we are not connected to presence, what do we want after getting everything we wanted?

More of what we just got.

# GET IN THE GAME

I am always amazed how many comments there are on social media platforms. In the tech industry, they refer to this as "engagement rates," how much people comment or like, either positively or negatively, a piece of content. And in our world today, engage we do! Video tends to produce more than text, but every day hundreds of millions of people spend countless hours commenting on what others are creating.

Now, does anyone read these comments? I have no idea. What I do know is that getting us to comment primarily helps the platform companies, not us. I have seen comments several pages long, all expressing a particular opinion the person wants heard. The person hates the video or post or loves the content. Then other people (and sometimes bots) comment as well, and hours a day are spent sharing opinions on content that someone else created.

**And it helps to ask, "Is this really how we want to spend our time?"**

It is as if we are on the sideline, watching a basketball game, and we keep yelling, "Pass the ball to this person!" or " Shoot the ball this way!" We are in the stands, not in the game, often thinking we know best. Of course, watching sports can be fun, but in the game of life, why don't we actually get in the game, instead of spending our time on the sidelines?

You could spend hours a day commenting and critiquing what other people post, or you could spend that same time developing your own content or creating what most inspires you. Why not do something? Put yourself out there. Let yourself be seen and critiqued. Step out from behind the curtain.

The game is playing, people are making moves. New podcasts, non-profits, and businesses are getting started, people are going for hikes, bike rides, connecting with nature … Do you want to spend your life commenting online about what everyone else is doing, or do you want to actually get in the game?

I am not saying commenting on other people's content is completely useless, but it helps to ask, is this really how we all want to spend our time?

**To be in the game and create rather than critique others is much more vulnerable … and valuable.**

It is very safe to be on the sidelines, thinking we know best and we could do so much better than those playing.

What will happen if we take that leap? We won't know until we do it. But we will likely face criticism, experience failure, encounter moments of humiliation, and more, but at least we are in the game! And as we put ourselves out there, we improve. We learn the game by playing it, not by critiquing others from the sidelines.

We live at a time when we can spend all our waking hours watching other people live their lives. "Look, Joe Rogan just took a cold plunge." Or "Oh my goodness, Taylor Swift just performed in Tokyo," or whatever. Whose life do we want to live? Someone else's or our own? For some people this could be a conscious choice, "I would rather not live my life, and instead watch other people live their lives."

Ok, that is fine, but for those who want to experience life in its many forms—love and loss, adventure and excitement, challenges and accomplishments—it helps to stop commenting and watching other people live, and get in the game ourselves. Join the game, help create the world you want to see.

So when there is the option, get in the game. Do something that requires feeling vulnerable. Move forward what matters most to you. All we have is our own life. We might as well live it rather than watch other people live theirs.

There are a few undeniable truths:

1. Our heartbeats in this life are limited.

2. We almost never know when our last heartbeat will come.

So, what are we waiting for?

# THE INVITATION OF CRITICISM

There's a story about a man who struggled with receiving criticism. In an effort to help, his teacher gave him an unusual task: "For one month, every time someone criticizes you, you have to give them $5."

At first, the idea seemed absurd, but he decided to follow his teacher's advice.

Sure enough, almost every day a friend, colleague, or family member said something critical, and he handed them $5. Naturally, this left the other person confused, but most were happy to take the money. He walked around with a lot of $5 bills in his pocket.

Finally, the month ended, and soon after a friend criticized him. Instead of recoiling and taking it personally as he did before the month-long exercise, he started laughing.

"Why are you laughing?" his friend inquired. "I just criticized you."

"Well," he said, "Previously I had to pay for this, and now I get to be criticized for free."

**The criticism directed toward him did not change, but his relationship to it did.**

Of course, there is criticism that is harsh and unkind, and it serves no purpose other than to harm. This type of criticism is not something we should cultivate, either in our interactions with others or in ourselves. For children especially, constant criticism can be devastating. At the same time, if we do anything worthwhile, we will be criticized. No great political leader, no business leader, no religious or spiritual leader, ever created needed change without immense criticism.

If we want to do anything meaningful in the world, many people will not like it. As we follow what matters to us, some people will come with us, and some will not; some people will be added, and some people will be subtracted. Some people will love us, and some will hate us.

And if we know what matters, our direction does not change. We create and do our work from a deeper source.

When we are deeply connected to that inner source, our work and actions are infused with clarity, creativity, and purpose. We no longer strive for approval or shield ourselves from criticism. Criticism still hurts but it does not sway us. We become channels for something greater, making our work not only more powerful but also more meaningful.

That terrible idea you can't believe came to mind—notice it, and let it go.

That brilliant idea you've been seeking for some time—notice it, and let it go.

When you no longer cling to or need anything from the great idea, consider following it.

# LEARNING FROM HUMILIATION

**P**eople with the greatest passion and courage often share one quality: they have experienced immense humiliation. They know failure. Life has not been easy, and the lack of "easiness" has strengthened them. It has given them resilience.

Of course, too much pain without love and support can be devastating. The intensity of the struggle can lead people to withdraw and isolate themselves. While pain and struggle are essential for growth, too much can become overwhelming.

This is one of the greatest challenges of parents: to know when to help their kids and when to let them suffer and develop resilience. The latter is often the hardest. Children need love and support, but at a certain age, they also need to know struggle and failure. There is a time to comfort and kiss the scraped knee, and there is a time to let them work it out. At times, our *help* does not help.

This is often why kids from wealthy families have such a hard time. They have not known hardship, it has not been a part of their life. They have not witnessed their parents failing, they have not done jobs like cleaning toilets or washing dishes where other people may view them as a failure. They have lived as "one of the successful people." This is what they know.

**"If you want to be successful, I would encourage you to grow a tolerance for failure." — Jensen Huang, CEO Nvidia**

So when failure and humiliation come during a business or life endeavor, which it will, it can be too much for them. Asking someone on a date, trying to connect with a new friend, learning a new language, or starting a dance class ... It is generally a lot of failing.

When failure comes, we have a choice: recoil, play it safe, avoid trying new things, stay home, watch the news, scroll social media, and find a vocation that is easy and safe .... Or we can experience failure and humiliation, and know that it is not who we are. We are with the experience, as best we can, and we know it is temporary. We allow it to grow us. We keep moving forward, if that is what calls us.

Of course, there will be setbacks and those are hard. When we go home for Thanksgiving or to a party, for example, and people ask us how we are, we can say with openness and confidence, "I started a project that by most accounts failed. It has been hard, but I am learning a lot about myself and looking for my next effort." That has power. We are a human being doing his or her best, and learning along the way. We can stand proudly in this.

We learn through trying, failing, then trying again ... enjoying as many of the moments along the way as possible. We do not want to fail, but we know who we are and stay true to that.

Sometimes, the best way to move a project forward is to become comfortable with it not moving forward.

We can notice the part of us that believes we will be a "better" person if it succeeds—or a "worse" person if it doesn't.

When that is cleared away, the effort can move forward. We can go all in.

# SOMEBODYNESS AND THE CREATIVE

I once worked for a celebrity, who at the time was one of the most popular actors in the country. When I started the job, I was struck by how quickly people responded to my emails and calls. Simply mentioning that I was reaching out on behalf of the celebrity triggered an almost instant reaction. It was a revealing lesson in how fame can, for better or worse, open doors and command attention with surprising ease.

After a year or so on the job, I became quite accustomed to this. Of course, I knew people were not responding because of me, but because of my relationship with this certain "somebody." I accepted it nonetheless. I was a "somebody" by association, and I would take that.

However, once I left the position, I was pained to discover that my somebodyness went with it. I was now just a regular person, and it sucked. My phone calls and emails were not answered with near the speed or consistency. I was now, for all intents and purposes, a "nobody." It was as if I had been robbed of my somebodyness by a thief in the night.

Today, there are various ways we measure "somebodyness": our job title, how we dress, the car we drive, and more often the number of followers on social media. We think, "the me that I think I am has status." And in this status, there are people above us and below us, and we think the journey of life is to move up the ranks. This is very much a tenet of the capitalist system, and it has both benefits and drawbacks.

When I shifted jobs, I first pleaded to the universe, "You have made a mistake. I am a somebody, not a nobody. Please correct this at once!" I brainstormed ways I might recover my somebodyness: "Is there another celebrity I could work for? How do I become important again?"

After some pain and suffering, I eventually realized that I had another choice: instead of trying to recreate the past, I could let it go. I could let the past be history. I saw that it was not any past event that was causing me suffering; it was the fact that I was bringing my past into my present. My attachment to "how things used to be" was making my present excruciatingly painful, and limiting my ability to respond creatively to my new conditions.

**The answer, I realized, was to let my past be history.**

This often plays out in any new effort. If we have failed in the past, we tend to think, "What if I fail again? I will be viewed even more negatively by others. I will lose what little status I have." And if we have succeeded in the past, we tend to think, "People think so well of me now. Do I really want to risk all this and try something new?" This is why so-called "successful" people have such a hard time risking both money and reputation.

In the end, it doesn't matter how many successes or failures we've experienced, how many businesses we created that have thrived or gone bankrupt, how many schools rejected or celebrated us, or how many jobs turned us away or embraced us. What truly matters is how fresh, open, and alive we are in this moment. This is where life is meant to be lived—right here, right now. Because, when we really reflect on it, the past is history, while the present is right here.

Of course, it matters on a practical level whether our efforts succeed. As a friend wisely told me, "Not all your efforts need to make money, but something you do needs to make money!" And the more we can let the past be history, the more freshly and creatively we can meet the opportunities in front of us. The best chance of responding creatively is found not by trying to recreate the past, but by letting it go, and creating from this moment right now.

When we let go, we are more open to create. We don't need to carry our past.

**What part of your past, be it what you might label as a success or a failure, are you still carrying?**

"Let go or be dragged."

— Zen Proverb

# ATTACHMENT AND NON-ATTACHMENT

"What an asshole," I remember thinking to myself. "How could you deny my request? I really needed you."

In the early days of creating my company, Wisdom 2.0, I worked hard to make it a success. I was starting with very little money and connections. The focus of the event was to gather the leaders in tech and in wisdom to explore how we can create a more livable world. While I knew people from the wisdom community, I really needed the tech leaders to show up.

When these leaders said no, as they often did, my initial reaction was the above.

I often thought, "If I can just get *this* particular person [fill in the blank with one of the most popular tech leaders of that time] everything will work out." People would then register for the event, other speakers would sign on, and I would make money.

As many of these tech leaders were extremely busy and did not know me, I got a "no" again and again and again. It was hard. I felt rejected, and often took it personally.

In that response, you could say, I was "attached" to a particular outcome. I blamed the tech leaders for my inability to grow the conference. I thought I needed them, and they were holding back my success. I resented them for giving me the answer they did, and they could likely feel it in my reaction.

**"What I learned from all my years in the media is to use every moment as a teachable moment." — Arianna Huffington**

Sometimes *no* is the teaching moment, and it is better to be graceful and practice what the Buddha called "non-attachment." Instead of complaining about the moment, we look for the teaching in it. We realize how much we do not know and how little we can actually control.

A more non-attached or graceful response is: "Thank you for getting back to me. I can only imagine how much you have to juggle, and understand you can only say yes to so many requests. I will circle back around next year. Maybe the stars will better align then. Wishing you well."

In the latter I am communicating, "I am committed to what I am doing. I will keep moving forward as best I can." I am not blaming them. In fact, I am building the connection.

I'm not suggesting we repress or deny the frustration that arises when we face a setback. Often, that frustration needs to run its course. Maybe you need to scream, punch some pillows, or vent to a friend for a while. Do what you need to do.

Then at some point, we realize that we have no idea what another person is going through. Perhaps they are dealing with a serious illness, a family member's struggles, or any number of other challenges. All we know is that, in that moment, the answer is "No." We can choose to respond with resentment or grace, with attachment or non-attachment. Attachment says, "It must not be this way. I will fight the current reality to make it how I wish." Non-attachment takes the approach, "This moment is what it is. I am not sure why or how it came to be, but here it is. I will respond as best as I can."

I see non-attachment as a great business and life skill. There are always no's in life, always setbacks, and these can either destroy or build relationships. One tech leader who has become a friend of mine turned down my request for four years, and said yes the fifth year. That was the time needed. Who am I to know how long something needs to take?

Of course, some efforts are more time sensitive, and they fail without the right support at the right time. Still, the best we can do is the best we can do: we practice getting disappointed, we let it run through us, and we stay the course (if we are so called to do so) and we prioritize the relationship (if it matters to us).

I have learned that rejection can be a great opportunity. And the response we choose in those moments says a lot about who we are and what we value. It can make us stronger or weaker, can build or break bonds depending on how we respond.

We get to plant a seed in those moments, and those seeds will determine future moments.

Attachment says, **"This moment should be different than it is."**

Non-attachment, or acceptance, says, **"This moment is as it is—and now I get to give the gift of my response."**

# THE ENDING

O f course, there is really no ending. Even when we die, our body continues on, decaying and feeding other life. Life simply takes a new form. All we really know is this: right now, here we are.

I started this book writing about the power of thoughts and stories, sharing lessons from my journey, the stumbles and the insights. The specifics of my life shouldn't mean much to you, they're my experiences. But if you apply any of this to your own life, that's wonderful, and that is the point. You have your own lessons that life has taught you, and these are far more important than any book, podcast, or movie. My hope is that this book has helped you reflect on your own lessons, your own wisdom.

The real point is to play, to learn, and to recognize the deeper intelligence in our lives. Every day brings opportunities for growth. Every moment has teachings. We don't need to go to a far-off place, visit a meditation center, or take psychedelics. We can do this if we are called, but nothing is mandatory. Life is always here, nature is always here, and our deeper nature is always here. We can connect to it at any moment.

In this sense, nothing truly ends, this is just another moment. If something resonated with you in this book, it's because you are that depth. No author, teacher, or guru is the point. The point is you. Life is everywhere, and it teaches us wherever we go. And you have everything you need inside you.

Of course, we often forget, and will likely forget again and again. We get triggered, we get lost in our stories, and we act and speak without awareness. It is part of life. But then we come back, and here we are again right now, in this moment. Life is incredibly forgiving. We get a new chance to live, each breath, each moment. We are each a part of a human experiment called to expand, to look deeply at who we really are and what really matters, and to live from that place.

To me, the only way we can truly address the challenges in front of us as a species, be it climate change or advancements in AI, is to access an intelligence inside us that is so much wiser and more powerful than any new intelligence we can create. It is not so much that "we" recognize this intelligence, it is more that "intelligence recognizes intelligence," "awareness recognizes awareness." We are it.

As such, the real teacher isn't this book, it's life itself. And to the degree we are connected to this moment, to life, the path becomes more clear.

We sense an inner guidance. Something else is here with us. We are not alone.

# ABOUT THE AUTHOR

**Soren Gordhamer** works and plays at the center of Ancient Wisdom and Modern Life.

He is the founder of Wisdom 2.0 and the Wisdom & AI Summit, as well as a partner at Wisdom Ventures.

You can join the free Wisdom 2.0 community via substack at www.wisdom2events.substack.com

See events, offerings and more at www.sorengordhamer.net